Plant Tissue Culture
Techniques and Experiments

T0282371

Plant Tissue Culture
Techniques and Experiments

Third edition

Roberta H. Smith

Emeritus Professor

Department of Horticulture
Vegetable Crops Improvement Center
Texas A&M University
College Station, Texas

ELSEVIER

AMSTERDAM • BOSTON • HEIDELBERG • LONDON
NEW YORK • OXFORD • PARIS • SAN DIEGO
SAN FRANCISCO • SINGAPORE • SYDNEY • TOKYO

Academic Press is an imprint of Elsevier

Academic Press is an imprint of Elsevier
32 Jamestown Road, London NW1 7BY, UK
225 Wyman Street, Waltham, MA 02451, USA
525 B Street, Suite 1800, San Diego, CA 92101-4495, USA

First edition 1996
Second edition 2006

Notice
No responsibility is assumed by the publisher for any injury and/or damage to persons or prop-
erty as a matter of products liability, negligence or otherwise, or from any use or operation of
any methods, products, instructions or ideas contained in the material herein. Because of rapid
advances in the medical sciences, in particular, independent verification of diagnoses and drug
dosages should be made.

British Library Cataloguing-in-Publication Data
A catalogue record for this book is available from the British Library

Library of Congress Cataloging-in-Publication Data
A catalog record for this book is available from the Library of Congress

ISBN: 978-0-12-415920-4

For information on all Academic Press publications
visit our website at elsevierdirect.com

Printed and bound in United States of America

12 13 14 15 16   10 9 8 7 6 5 4 3 2 1

Working together to grow
libraries in developing countries

www.elsevier.com | www.bookaid.org | www.sabre.org

ELSEVIER   BOOK AID
             International   Sabre Foundation

Contents

This manual resulted from the need for plant tissue culture laboratory exercises that demonstrate major concepts and that use plant material that is available year round. The strategy in developing this manual was to devise exercises that do not require maintenance of an extensive collection of plant materials, yet give the student the opportunity to work on a wide array of plant materials.

The students who have used these exercises range from high school (science fair and 4-H projects) to undergraduate, graduate and post-doctoral levels. The manual is predominantly directed at students who are in upper-level college or university classes and who have taken courses in chemistry, plant anatomy, and plant physiology.

Before starting the exercises, students should examine Chapters 2 through 5, which deal with the setup of a tissue culture laboratory, media preparation, explants, aseptic technique, and contamination. The information in these chapters will be needed in the exercises that follow.

The brief introduction to each chapter is not intended to be a review of the chapter's topic but rather to complement lecture discussions of the topic. In this revised edition, Dr. Trevor Thorpe has contributed a chapter on the history of plant cell culture. Dr. Brent McCown contributed a chapter on woody trees and shrubs. Dr. Sunghun Park, Jungeun Kim Park, and James E. Craven have contributed a chapter on protoplast isolation and fusion. Jungeun Kim Park, Dr. Sunghun Park, and Qingyu Wu contributed a chapter on *Agrobacterium*-mediated transformation of plants.

In many instances, plant material initiated in one exercise is used in subsequent exercises. Refer to Scheduling and Interrelationships of Exercises to obtain information on the time required to complete the exercises and how they relate to one another.

All of the exercises have been successfully accomplished for at least 15 semesters. Tissue culture, however, is still sometimes more art than science, and variation in individual exercises can be expected.

Roberta H. Smith

Acknowledgments

I acknowledge the many teaching assistants who helped in developing some of these exercises: Daniel Caulkins, Cheryl Knox, John Finer, Richard Norris, Ann Reilley-Panella, Ricardo Diquez, Eugenio Ulian, Shelly Gore, Sara Perez-Ramos, Jeffery Callin, Greg Peterson, Sunghun Park, Maria Salas, Metinee Srivantanakul, and Cecilia Zapata. Additionally, all the students who have taken this course since 1979 have been instrumental in developing and improving these exercises.

The contributions of Dr. Trevor Thorpe with the chapter on the history of plant cell culture, Dr. Brent McCown with a chapter on woody trees and shrubs, and Dr. Sunghun Park, Jungeun Kim Park, James E. Craven, and Qingyu Wu with the chapters on protoplast isolation and fusion and *Agrobacterium*-mediated transformation are tremendously appreciated.

Last, I thank my husband, Jim, daughter, Cristine, and son, Will, their spouses, Gaylon and Trudy, and my grandchildren, Claire Jean, William, Clayton, Wyatt, and Grant. Especially the grandchildren for taking their naps while I had high speed internet at their homes to access library databases to update this edition.

Scheduling and Interrelationships of Exercises

   I. Aseptic Germination of Seed (Chapter 4)
      Carrot: 1–2 weeks; Cotton, Sunflower: 1 week
      a. Callus Induction (Chapter 6): 6 weeks
      Broccoli, Lemon 6–8 weeks
      Carrot: 2 subcultures, 6 weeks each = 3 months
        1. Salt Selection *in Vitro* (Chapter 6): 4 weeks
        2. Suspension Culture (Chapter 7): 2 weeks
          Carrot
          a. Somatic Embryogenesis (Chapter 7): 3–4 weeks
          b. Explant Orientation (Chapter 6): 6 weeks
            Cotton: 2 subcultures, 6 weeks each
        1. Protoplast (Chapter 13): 2 days
        2. Cellular Variation (Chapter 6): 4 weeks
        3. Growth Curves (Chapter 6): 6 weeks
  II. Tobacco Seed Germination (Chapter 6): 3 weeks
      a. Callus Induction (Chapter 6): 2 subcultures, 6 weeks each
 III. Establishment of Competent Cereal Cell Cultures (Chapter 6): 2–3 weeks
      a. Rice Subculture (Chapter 7): 3 weeks
        1. Plant Regeneration: 4–6 weeks
 IV. Potato Shoot Initiation (Chapter 7): 6 weeks
      a. Potato Tuberization (Chapter 7): 4–6 weeks
  V. Douglas Fir Seed Germination (Chapter 4): 2–4 weeks
      a. Primary Morphogenesis (Chapter 7): 4 weeks
 VI. Petunia/Tobacco Leaf Disk Transformation (Chapter 14): 6 weeks
VII. Petunia Shoot Apex Transformation (Chapter 14): 4–6 weeks
VIII. Solitary Exercises
      a. Bulb Scale Dormancy (Chapter 7): 6–8 weeks
      b. Datura Anther Culture (Chapter 9): 4–8 weeks; 10 weeks to obtain flowering plants
      c. African Violet Anther Culture (Chapter 9): 7–8 weeks
      d. Tobacco Anther Culture (Chapter 9): 7–8 weeks; 2–3 months to obtain flowering plants
      e. Corn Embryo Culture (Chapter 10): 72 hr
      f. Crabapple and Pear Embryo Culture (Chapter 10): 2–3 weeks
      g. Shoot Apical Meristem (Chapter 11): 4–6 weeks

h. Diffenbachia Meristem (Chapter 11): 4–6 weeks
i. Garlic Propagation (Chapter 11): 4 weeks
j. Boston Fern Propagation (Chapter 12)
   Stage I: 6–8 weeks
   Stage II: 4–6 weeks
   Stage III: 2–3 weeks
k. Staghorn Fern Propagation (Chapter 12)
   Stage I: 2–3 weeks
   Stage II: 6 weeks
   Stage III: 4–6 weeks
l. Ficus Propagation (Chapter 12)
   Stage I: 4–6 weeks
   Stage II: 4–6 weeks
   Stage III: 4 weeks
m. Kalanchoe Propagation, Stages I & II (Chapter 12): 4 weeks
n. African Violet, Stages I & II (Chapter 12): 4 weeks
o. Pitcher Plant, Stages I & II (Chapter 12): 6 weeks
p. Cactus Propagation (Chapter 12)
   Stage I: 4–6 weeks
   Stage II: 4–6 weeks
   Stage III: 8 weeks
q. Rhododendrons and Azaleas (Chapter 8): 4–6 weeks
r. Birch Trees (Chapter 8): 2 weeks seed germination: 4–6 weeks
s. White Cedar (Chapter 8): 4–6 weeks
t. Roses (Chapter 8): 4–6 weeks

Chapter 1

History of Plant Cell Culture

Trevor A. Thorpe
The University of Calgary

Chapter Outline

INTRODUCTION

Plant cell/tissue culture, also referred to as *in vitro,* axenic, or sterile culture, is an important tool in both basic and applied studies as well as in commercial application (see Thorpe, 1990, 2007 and Stasolla & Thorpe 2011). Although Street (1977) has recommended a more restricted use of the term, plant tissue culture is generally used for the aseptic culture of cells, tissues, organs, and their components under defined physical and chemical conditions *in vitro.* Perhaps the earliest step toward plant tissue culture was made by Henri-Louis Duhumel du Monceau in 1756, who, during his pioneering studies on wound-healing in plants, observed callus formation (Gautheret, 1985). Extensive microscopic studies led to the independent and almost simultaneous development of the cell theory by Schleiden (1838) and Schwann (1839). This theory holds that the cell is the unit of structure and function in an organism and therefore capable of autonomy. This idea was tested by several researchers, but the work of Vöchting (1878) on callus formation and on the limits to divisibility of plant segments was perhaps the most important. He showed that the upper part of a stem segment always produced buds and the lower end callus or

Plant Tissue Culture. Third Edition. DOI: 10.1016/B978-0-12-415920-4.00001-3
1

roots independent of the size until a very thin segment was reached. He demonstrated polar development and recognized that it was a function of the cells and their location relative to the cut ends.

The theoretical basis for plant tissue culture was proposed by Gottlieb Haberlandt in his address to the German Academy of Science in 1902 on his experiments on the culture of single cells (Haberlandt, 1902). He opined that to "my knowledge, no systematically organized attempts to culture isolated vegetative cells from higher plants have been made. Yet the results of such culture experiments should give some interesting insight to the properties and potentialities which the cell as an elementary organism possesses. Moreover, it would provide information about the inter-relationships and complementary influences to which cells within a multicellular whole organism are exposed" (from the English translation by Krikorian & Berquam, 1969). He experimented with isolated photosynthetic leaf cells and other functionally differentiated cells and was unsuccessful, but nevertheless he predicted that "one could successfully cultivate artificial embryos from vegetative cells." He thus clearly established the concept of totipotency, and further indicated that "the technique of cultivating isolated plant cells in nutrient solution permits the investigation of important problems from a new experimental approach." On the basis of that 1902 address and his pioneering experimentation before and later, Haberlandt is justifiably recognized as the father of plant tissue culture. Greater detail on the early pioneering events in plant tissue culture can be found in White (1963), Bhojwani and Razdan (1983), and Gautheret (1985).

THE EARLY YEARS

Using a different approach Kotte (1922), a student of Haberlandt, and Robbins (1922) succeeded in culturing isolated root tips. This approach, of using explants with meristematic cells, led to the successful and indefinite culture of tomato root tips by White (1934a). Further studies allowed for root culture on a completely defined medium. Such root cultures were used initially for viral studies and later as a major tool for physiological studies (Street, 1969). Success was also achieved with bud cultures by Loo (1945) and Ball (1946).

Embryo culture also had its beginning early in the nineteenth century, when Hannig in 1904 successfully cultured cruciferous embryos and Brown in 1906 barley embryos (Monnier, 1995). This was followed by the successful rescue of embryos from nonviable seeds of a cross between *Linum perenne* × *L. austriacum* (Laibach, 1929). Tukey (1934) was able to allow for full embryo development in some early-ripening species of fruit trees, thus providing one of the earliest applications of *in vitro* culture. The phenomenon of precocious germination was also encountered (LaRue, 1936).

The first true plant tissue cultures were obtained by Gautheret (1934, 1935) from cambial tissue of *Acer pseudoplatanus*. He also obtained success with similar explants of *Ulmus campestre*, *Robinia pseudoacacia*, and *Salix capraea* using agar-solidified medium of Knop's solution, glucose, and cysteine hydro-chloride. Later, the availability of indole acetic acid and the addition of B vitamins allowed for the more or less simultaneous demonstrations by Gautheret (1939) and Nobécourt (1939a) with carrot root tissues and White (1939a) with tumor tissue of a *Nicotiana glauca* × *N. langsdorffii* hybrid, which did not require auxin, that tissues could be continuously grown in culture and even made to differentiate roots and shoots (Nobécourt, 1939b; White, 1939b). However, all of the initial explants used by these pioneers included meristematic tissue. Nevertheless, these findings set the stage for the dramatic increase in the use of *in vitro* cultures in the subsequent decades.

THE ERA OF TECHNIQUES DEVELOPMENT

The 1940s, 1950s, and 1960s proved an exciting time for the development of new techniques and the improvement of those already available. The application of coconut water (often incorrectly stated as coconut milk) by Van Overbeek *et al.* (1941) allowed for the culture of young embryos and other recalcitrant tissues, including monocots. As well, callus cultures of numerous species, including a variety of woody and herbaceous dicots and gymnosperms as well as crown gall tissues, were established (see Gautheret, 1985). Also at this time, it was recognized that cells in culture underwent a variety of changes, including loss of sensitivity to applied auxin or habituation (Gautheret, 1942, 1955) as well as variability of meristems formed from callus (Gautheret, 1955; Nobécourt, 1955). Nevertheless, it was during this period that most of the *in vitro* techniques used today were largely developed.

Studies by Skoog and his associates showed that the addition of adenine and high levels of phosphate allowed nonmeristematic pith tissues to be cultured and to produce shoots and roots, but only in the presence of vascular tissue (Skoog & Tsui, 1948). Further studies using nucleic acids led to the discovery of the first cytokinin (kinetin) as the breakdown product of herring sperm DNA (Miller *et al.*, 1955). The availability of kinetin further increased the number of species that could be cultured indefinitely, but perhaps most importantly, led to the recognition that the exogenous balance of auxin and kinetin in the medium influenced the morphogenic fate of tobacco callus (Skoog & Miller, 1957). A relative high level of auxin to kinetin favored rooting, the reverse led to shoot formation, and intermediate levels to the proliferation of callus or wound paren-chyma tissue. This morphogenic model has been shown to operate in numerous species (Evans *et al.*, 1981). Native cytokinins were subsequently discovered in several tissues, including coconut water (Letham, 1974). In addition to the for-mation of unipolar shoot buds and roots, the formation of bipolar somatic embryos (carrot) were first reported independently by Reinert (1958, 1959) and Steward *et al.* (1958).

The culture of single cells (and small cell clumps) was achieved by shaking callus cultures of *Tagetes erecta* and tobacco and subsequently placing them on filter paper resting on well-established callus, giving rise to the so-called nurse culture (Muir *et al.*, 1954, 1958). Later, single cells could be grown in medium in which tissues had already been grown, i.e., conditioned medium (Jones *et al.*, 1960). As well, Bergmann (1959) incorporated single cells in a 1-mm layer of solidified medium where some cell colonies were formed. This technique is widely used for cloning cells and in protoplast culture (Bhojwani & Razdan, 1983). Kohlenbach (1959) finally succeeded in the culture of mechanically isolated mature differentiated mesophyll cells of *Macleaya cordata* and later induced somatic embryos from callus (Kohlenbach, 1966). The first large-scale culture of plant cells was reported by Tulecke and Nickell (1959), who grew cell suspensions of *Ginkgo*, holly, *Lolium*, and rose in simple sparged 20-liter carboys. Utilizing coconut water as an additive to fresh medium, instead of using conditioned medium, Vasil and Hildebrandt (1965) finally realized Haberlandt's dream of producing a whole plant (tobacco) from a single cell, thus demonstrating the totipotency of plant cells.

The earliest nutrient media used for growing plant tissues *in vitro* were based on the nutrient formulations for whole plants, for which they were many (White, 1963); but Knop's solution and that of Uspenski and Uspenskia were used the most and provided less than 200 mg/liter of total salts. Heller (1953), based on studies with carrot and Virginia creeper tissues, increased the concentration of salts twofold, and Nitsch and Nitsch (1956) further increased the salt concentration to ca 4 g/liter, based on their work with Jerusalem artichoke. However, these changes did not provide optimum growth for tissues, and complex addenda, such as yeast extract, protein hydrolysates, and coconut water, were frequently required. In a different approach based on an examination of the ash of tobacco callus, Murashige and Skoog (1962) developed a new medium. The concentration of some salts were 25 times that of Knop's solution. In particular, the level of NO_3^- and NH_4^+ were very high and the array of micronutrients were increased. This formulation allowed for a further increase in the number of plant species that could be cultured, many of them using only a defined medium consisting of macro- and micronutrients, a carbon source, reduced nitrogen, B vitamins, and growth regulators (Gamborg *et al.*, 1976).

Ball (1946) successfully produced plantlets by culturing shoot tips with a couple of primordia of *Lupinus* and *Tropaeolum*, but the importance of this finding was not recognized until Morel (1960), using this approach to obtain virus-free orchids, realized its potential for clonal propagation. The potential was rapidly exploited, particularly with ornamentals (Murashige, 1974). Early studies by White (1934b) showed that cultured root tips were free of viruses. Later Limmaset and Cornuet (1949) observed that the virus titer in the shoot meristem was very low. This was confirmed when virus-free *Dahlia* plants were obtained from infected plants by culturing their shoot tips (Morel & Martin, 1952). Virus elimination was possible because vascular tissue, in which the viruses move, do not extend into the root or shoot apex. The method was further refined by Quack (1961) and is now routinely used.

Techniques for *in vitro* culture of floral and seed parts were developed during this period. The first attempt at ovary culture was by LaRue (1942), who obtained limited growth of ovaries accompanied by rooting of pedicels in several species. Compared to studies with embryos, successful ovule culture is very limited. Studies with both ovaries and ovules have been geared mainly to an understanding of factors regulating embryo and fruit development (Rangan, 1982). The first continuously growing tissue cultures from an endosperm were from immature maize (LaRue, 1949); later, plantlet regeneration via organogenesis was achieved in *Exocarpus cupressiformis* (Johri & Bhojwani, 1965).

In vitro pollination and fertilization was pioneered by Kanta *et al.* (1962) using *Papaver somniferum*. The approach involves culturing excised ovules and pollen grains together in the same medium and has been used to produce interspecific and intergeneric hybrids (Zenkteler *et al.*, 1975). Earlier, Tuleke (1953) obtained cell colonies from *Ginkgo* pollen grains in culture, and Yamada *et al.* (1963) obtained haploid callus from whole anthers of *Tradescantia reflexa*. However, it was the finding of Guha and Maheshwari (1964, 1966) that haploid plants could be obtained from cultured anthers of *Datura innoxia* that opened the new area of androgenesis. Haploid plants of tobacco were also obtained by Bourgin and Nitsch (1967), thus confirming the totipotency of pollen grains.

Plant protoplasts or cells without cell walls were first mechanically isolated from plasmolyzed tissues well over 100 years ago by Klercker in 1892, and the first fusion was achieved by Küster in 1909 (Gautheret, 1985). Nevertheless, this remained an unexplored technology until the use of a fungal cellulase by Cocking (1960) ushered in a new era. The commercial availability of cell-wall-degrading enzymes led to their wide use and the development of protoplast technology in the 1970s. The first demonstration of the totipotency of protoplasts was by Takebe *et al.* (1971), who obtained tobacco plants from mesophyll protoplasts. This was followed by the regeneration of the first interspecific hybrid plants (*Nicotiana glauca* × *Nicotiana langsdorffii*) by Carlson *et al.* (1972).

Braun (1941) showed that *Agrobacterium tumefaciens* could induce tumors in sunflower, not only at the inoculated sites, but at distant points. These secondary tumors were free of bacteria and their cells could be cultured without auxin (Braun & White, 1943). Further experiments showed that crown gall tissues, free of bacteria, contained a tumor-inducing principle (TIP), which was probably a macromolecule (Braun, 1950). The nature of the TIP was worked out in the 1970s (Zaenen *et al.*, 1974), but Braun's work served as the foundation for *Agrobacterium*-based transformation. It should also be noted that the finding by Ledoux (1965) that plant cells could take up and integrate DNA remained controversial for over a decade.

THE RECENT PAST

Based on the availability of the various *in vitro* techniques discussed above, it is not surprising that, starting in the mid-1960s, there was a dramatic increase in their application to various problems in basic biology, agriculture, horticulture, and forestry through the 1970s and 1980s. These applications can be divided conveniently into five broad areas, namely: (a) cell behavior, (b) plant modification and improvement, (c) pathogen-free plants and germplasm storage, (d) clonal propagation, and (e) product formation (Thorpe, 1990). Detailed information on the approaches used can be gleaned from Bhojwani and Razdan (1983), Vasil (1984), Vasil and Thorpe (1994), and Stasolla and Thorpe (2011), among several sources.

Cell Behavior

Included under this heading are studies dealing with cytology, nutrition, and primary and secondary metabolism as well as morphogenesis and pathology of cultured tissues (Thorpe, 1990). Studies on the structure and physiology of quiescent cells in explants, changes in cell structure associated with the induction of division in these explants, and the characteristics of developing callus and cultured cells and protoplasts have been carried out using light and electron microscopy (Yeoman & Street, 1977; Lindsey & Yeoman, 1985; Fowke 1986, 1987). Nuclear cytology studies have shown that endoreduplication, endomitosis, and nuclear fragmentation are common features of cultured cells (D'Amato, 1978; Nagl et al., 1985).

Nutrition was the earliest aspect of plant tissue culture investigated, as indicated earlier. Progress has been made in the culture of photoautotrophic cells (Yamada et al., 1978; Hüsemann, 1985). In vitro cultures, particularly cell suspensions, have become very useful in the study of both primary and secondary metabolism (Neumann et al., 1985). In addition to providing protoplasts from which intact and viable organelles were obtained for study (e.g., vacuoles; Leonard & Rayder, 1985), cell suspensions have been used to study the regulation of inorganic nitrogen and sulfur assimilation (Filner, 1978), carbohydrate metabolism (Fowler, 1978), and photosynthetic carbon metabolism (Bender et al., 1985; Herzbeck & Hüsemann, 1985), thus clearly showing the usefulness of cell cultures for elucidating pathway activity. Most of the work on secondary metabolism was related to the potential of cultured cells to form commercial products, but has also yielded basic biochemical information (Constabel & Vasil, 1987, 1988).

Morphogenesis or the origin of form is an area of research with which tissue culture has long been associated and one to which tissue culture has made significant contributions in terms of both fundamental knowledge and application (Thorpe, 1990). Xylogenesis or tracheary element formation has been used to study cytodifferentiation (Roberts, 1976; Phillips, 1980; Fukuda & Komamine, 1985). In particular the optimization of the Zinnia mesophyll single-cell system has dramatically improved our knowledge of this process. The classic findings of Skoog and Miller (1957) on the hormonal balance for organogenesis has continued to influence research on this topic, a concept supported more recently by transformation of cells with appropriately modified *Agrobacterium* T-DNA (Schell *et al.*, 1982; Schell, 1987). However, it is clear from the literature that several additional factors, including other growth-active substances, interact with auxin and cytokinin to bring about *de novo* organogenesis (Thorpe, 1980). In addition to bulky explants, such as cotyledons, hypocotyls, and callus (Thorpe, 1980), thin (superficial) cell layers (Tran Thanh Van & Trinh, 1978; Tran Thanh Van, 1980) have been used in traditional morphogenic studies, as well as to produce *de novo* organs and plantlets in hundreds of plant species (Murashige, 1974, 1979). Furthermore, physiological and biochemical studies on organogenesis have been carried out (Thorpe, 1980; Brown & Thorpe, 1986; Thompson & Thorpe, 1990). The third area of morphogenesis, somatic embryogenesis, also developed in this period and by the early 1980s over 130 species were reported to form bipolar structures (Ammirato, 1983; Thorpe, 1988). Successful culture was achieved with cereals, grasses, legumes, and conifers, previously considered to be recalcitrant groups. The development of a single-cell-to-embryo system in carrot (Normura & Komamine, 1985) allows for an in-depth study of the process.

Cell cultures have continued to play an important role in the study of plant–microbe interaction, not only in tumorigenesis (Butcher, 1977), but also on the biochemistry of virus multiplication (Rottier, 1978), phytotoxin action (Earle, 1978), and disease resistance, particularly as affected by phytoalexins (Miller & Maxwell, 1983). Without doubt the most important studies in this area dealt with *Agrobacteria*, and, although aimed mainly at plant improvement (see below), provided good fundamental information (Schell, 1987).

Plant Modification and Improvement

During this period *in vitro* methods were used increasingly as an adjunct to traditional breeding methods for the modification and improvement of plants. The technique of controlled *in vitro* pollination on the stigma, placenta, or ovule has been used for the production of interspecific and intergeneric hybrids, overcoming sexual self-incompatibility, and the induction of haploid plants (Yeung *et al.*, 1981; Zenkteler, 1984). Embryo, ovary, and ovule cultures have been used in overcoming embryo inviability, monoploid production in barley, and seed dormancy and related problems (Raghavan, 1980; Yeung *et al.*, 1981). In particular, embryo rescue has played a most important role in producing interspecific and intergeneric hybrids (Collins & Grosser, 1984).

By the early 1980s, androgenesis had been reported in some 171 species, of which many were important crop plants (Hu & Zeng, 1984). Gynogenesis was reported in some 15 species, in some of which androgenesis was not successful (San & Gelebart, 1986). The value of these haploids was that they could be used to detect mutations and for recovery of unique recombinants, since there is no masking of recessive alleles. As well, the production of double-haploids allowed for hybrid production and their integration into breeding programs.

Cell cultures have also played an important role in plant modification and improvement, as they offer advantages for isolation of variants (Flick, 1983). Although tissue culture-produced variants have been known since the 1940s, e.g., habituation, it was only in the 1970s that attempts were made to utilize them for plant improvement. This somaclonal variation is dependent on the natural variation in a population of cells, either preexisting or culture induced, and is usually observed in regenerated plantlets (Larkin & Scowcroft, 1981). The variation may be genetic or epigenetic and is not simple in origin (Larkin *et al.*, 1985; Scowcroft *et al.*, 1987), but the changes in the regenerated plantlets have potential agricultural and horticultural significance. It has also been possible to produce a wide spectrum of mutant cells in culture (Jacobs *et al.*, 1987). These include cells showing biochemical differences and antibiotic-, herbicide-, and stress-resistance. In addition, auxotrophs, autotrophs, and those with altered developmental systems have been selected in culture; usually the application of the selective agent in the presence of a mutagen is required. However, in only a few cases has it been possible to regenerate plants with the desired traits, e.g., herbicide-resistant tobacco (Hughes, 1983) and methyl tryptophan-resistant *Datura innoxia* (Ranch *et al.*, 1983).

By 1985 nearly 100 species of angiosperms could be regenerated from protoplasts (Binding, 1986). The ability to fuse plant protoplasts by chemical (e.g., with PEG) and physical (e.g., electrofusion) means allowed for production of somatic hybrid plants, the major problem being the ability to regenerate plants from the hybrid cells (Evans *et al.*, 1984; Schieder & Kohn, 1986). Protoplast fusion has been used to produce unique nuclear-cytoplasmic combinations. In one such example, *Brassica campestris* chloroplasts coding for atrazine resistance (obtained from protoplasts) were transferred into *Brassica napus* protoplasts with *Raphanus sativus* cytoplasm (which confers cytoplasmic male sterility from its mitochondria). The selected plants contained *B. napus* nuclei, chloroplasts from *B. campestris,* and mitochondria from *R. sativus*; had the desired traits in a *B. napus* phenotype; and could be used for hybrid seed production (Chetrit *et al.*, 1985). Unfortunately, only a few such examples exist.

Genetic modification of plants is being achieved by direct DNA transfer via vector-independent and vector-dependent means since the early 1980s. Vector-independent methods with protoplasts include electroporation (Potrykus *et al.*, 1985), liposome fusion (Deshayes *et al.*, 1985), and microinjection (Crossway *et al.*, 1986), as well as high-velocity microprojectile bombardment (biolistics) (Klein *et al.*, 1987). This latter method can be executed with cells, tissues, and organs. The use of *Agrobacterium* in vector-mediated transfer has progressed very rapidly since the first reports of stable transformation (DeBlock *et al.*, 1984; Horsch *et al.*, 1984). Although the early transformations utilized protoplasts, regenerable organs such as leaves, stems, and roots have been subsequently used (Gasser & Fraley, 1989; Uchimiya *et al.*, 1989). Much of the research activity utilizing these tools has focused on engineering important agricultural traits for the control of insects, weeds, and plant diseases.

Pathogen-Free Plants and Germplasm Storage

Although these two uses of *in vitro* technology may appear unrelated, a major use of pathogen-free plants is for germplasm storage and the movement of living material across international borders (Thorpe, 1990). The ability to rid plants of viruses, bacteria, and fungi by culturing meristem tips has been widely used since the 1960s. The approach is particularly needed for virus-infected material, as bactericidal and fungicidal agents can be used successfully in ridding plants of bacteria and fungi (Bhojwani & Razdan, 1983). Meristem-tip culture is often coupled with thermotherapy or chemotherapy for virus eradication (Kartha, 1981).

Traditionally, germplasm has been maintained as seed, but the ability to regenerate whole plants from somatic and gametic cells and shoot apices has led to their use for storage (Kartha, 1981; Bhojwani & Razdan, 1983). Three *in vitro* approaches have been developed, namely use of growth retarding compounds (e.g., maleic hydrazide, B995, and ABA; Dodds, 1989), low nonfreezing temperatures ($1–9°C$; Bhojwani & Razdan, 1983), and cryopreservation (Kartha, 1981). In this last approach, cell suspensions, shoot apices, asexual embryos, and young plantlets, after treatment with a cryoprotectant, are frozen and stored at the temperature of liquid nitrogen (ca. $−196°C$) (Kartha, 1981; Withers, 1985)

Clonal Propagation

The use of tissue culture technology for the vegetative propagation of plants is the most widely used application of the technology. It has been used with all classes of plants (Murashige, 1978; Conger, 1981), although some problems still need to be resolved, e.g., hyperhydricity and abberant plants. There are three ways by which micropropagation can be achieved. These are enhancing axillary bud-breaking, production of adventitious buds directly or indirectly via callus, and somatic embryogenesis directly or indirectly on explants (Murashige, 1974, 1978). Axillary bud-breaking produces the smallest number of plantlets, but they are generally genetically true-to-type, while somatic embryogenesis has the potential to produce the greatest number of plantlets but is induced in the lowest number of plant species. Commercially, numerous ornamentals are produced, mainly via axillary bud-breaking (Murashige, 1990). As well, there are lab-scale protocols for other classes of plants, including field and vegetable crops and fruit, plantation, and forest trees, but cost of production is often a limiting factor in their use commercially (Zimmerman, 1986).

Product Formation

Higher plants produce a large number of diverse organic chemicals, which are of pharmaceutical and industrial interest. The first attempt at the large-scale culture of plant cells for the production of pharmaceuticals took place in the 1950s at the

Charles Pfizer Company (U.S.). The failure of this effort limited research in this area in the U.S., but work in Germany and Japan, in particular, led to development so that by 1978 the industrial application of cell cultures was considered feasible (Zenk, 1978). Furthermore, by 1987 there were 30 cell culture systems that were better producers of secondary metabolites than the respective plants (Wink, 1987). Unfortunately, many of the economically important plant products are either not formed in sufficiently large quantities or not at all by plant cell cultures. Different approaches have been taken to enhance yields of secondary metabolites. These include cell-cloning and the repeated selection of high-yielding strains from heterogeneous cell populations (Zenk, 1978; Dougall, 1987) and by using ELISA and radioimmunoassay techniques (Kemp & Morgan, 1987). Another approach involves selection of mutant cell lines that overproduce the desired product (Widholm, (1987). As well, both abiotic factors, such as UV irridiation, exposure to heat or cold and salts of heavy metals, and biotic elicitors of plant and microbial origin, have been shown to enhance secondary product formation (Eilert, 1987; Kurz, 1988). Last, the use of immobilized cell technology has also been examined (Brodelius, 1985; Yeoman, 1987).

Central to the success of producing biologically active substances commercially is the capacity to grow cells on a large scale. This is being achieved using stirred tank reactor systems and a range of air-driven reactors (Fowler, 1987). For many systems, a two-stage (or two-phase) culture process has been tried (Beiderbeck & Knoop, 1987; Fowler, 1987). In the first stage, rapid cell growth and biomass accumulation are emphasized, while the second stage concentrates on product synthesis with minimal cell division or growth. However, by 1987 the naphthoquinone shikonin was the only commercially produced secondary metabolite from cell cultures (Fujita & Tabata, 1987).

THE PRESENT ERA

During the 1990s and the early twenty-first century continued expansion in the application of *in vitro* technologies to an increasing number of plant species has been observed. Tissue culture techniques are being used with all types of plants, including cereals and grasses (Vasil & Vasil, 1994), legumes (Davey *et al.,* 1994), vegetable crops (Reynolds, 1994), potato (Jones, 1994) and other root and tuber crops (Krikorian, 1994a), oilseeds (Palmer & Keller, 1994), temperate (Zimmerman & Swartz, 1994) and tropical (Grosser, 1994) fruits, plantation crops (Krikorian, 1994b), forest trees (Harry & Thorpe, 1994), and, of course, ornamentals (Debergh, 1994). As will be seen from these articles, the application of *in vitro* cell technology goes well beyond micropropagation and embraces all the *in vitro* approaches that are relevant or possible for the particular species and the problem(s) being addressed. However, only limited success has been achieved in exploiting somaclonal variation (Karp, 1994) or in the regeneration of useful plantlets from mutant cells (Dix, 1994); also, the early promise of protoplast technology remains largely unfulfilled (Feher & Dudits, 1994). Good progress is being made in extending cryopreservation technology for germ-plasm storage (Kartha & Engelmann, 1994). Progress is also being made in artificial seed technology (Redenbaugh, 1993).

Cell cultures have remained an important tool in the study of plant biology. Thus progress is being made in cell biology, for example, in studies of the cyto-skeleton (Kong *et al.*, 1998), on chromosomal changes in cultured cells (Kaeppler & Phillips, 1993), and in cell cycle studies (Komamine *et al.*, 1993; Trehin *et al.*, 1998). Better physiological and biochemical tools have allowed for a reexamina-tion of neoplastic growth in cell cultures during habituation and hyperhydricity and relate it to possible cancerous growth in plants (Gaspar, 1995). Cell cultures have remained an extremely important tool in the study of primary metabolism; for example, the use of cell suspensions to develop *in vitro* transcription systems (Suguira, 1997) or the regulation of carbohydrate metabolism in transgenics (Stitt & Sonnewald, 1995). The development of medicinal plant cell culture techniques has led to the identification of more than 80 enzymes of alkaloid biosynthesis (reviewed in Kutchan, 1998). Similar information arising from the use of cell cultures for molecular and biochemical studies on other areas of secondary metabolism is generating research activity on metabolic engineering of plant sec-ondary metabolite production (Verpoorte *et al.*, 1998).

Cell cultures remain an important tool in the study of morphogenesis, even though the present use of developmental mutants, particularly of *Arabidopsis*, is adding valuable information on plant development (e.g., see *The Plant Cell* (Special Issue), July, 1997). Molecular, physiological, and biochemical studies are allowing for in-depth understanding of cytodifferentiation, mainly tracheary element formation (Fukuda, 1997), organogenesis (Thorpe, 1993; Thompson & Thorpe, 1997), and somatic embryogenesis (Nomura & Komamine, 1995; Dudits *et al.*, 1995).

Advances in molecular biology allow for the genetic engineering of plants through the precise insertion of foreign genes from diverse biological systems. Three major breakthroughs have played major roles in the development of this transformation technology (Hinchee *et al.*, 1994). These are the development of shuttle vectors for harnessing the natural gene transfer capability of *Agrobacterium* (Fraley *et al.*, 1985), the methods to use these vectors for the direct transformation of regenerable explants obtained from plant organs (Horsch *et al.*, 1985), and the development of selectable markers (Cloutier & Landry, 1994). For species not amenable to *Agrobacterium*-mediated transformation, physical, chemical, and mechanical means are used to get the DNA into the cells. With these latter approaches, particularly biolistics, it is becoming possible to transform any plant species and genotype.

The initial wave of research in plant biotechnology has been driven mainly by the seed and agrichemical industries and has concentrated on "agronomic traits" of direct relevance to these industries, namely the control of insects, weeds, and plant diseases (Fraley, 1992). At present, over 100 species of plants have been genetically engineered, including nearly all the major dicotyledonous crops and an increasing number of monocotyledonous ones as well as some woody plants. Current research has led to routine gene transfer systems for most important crops. In addition, technical improvements are further increasing transformation efficiency, extending transformation to elite commercial germplasm and lowering transgenic plant production costs. The next wave in agricultural biotechnology is already in progress with biotechnological applications of interest to the food processing, speciality chemical, and pharmaceutical industries. Also see Datta (2007).

The current emphasis and importance of plant biotechnology can be gleaned from the IXth International Congress on Plant Tissue and Cell Culture held in Israel in June, 1998. The theme of the Congress was "Plant Biotechnology and In Vitro Biology in the 21st Century." This theme was developed through a scientific program which focused on the most important developments, both basic and applied, in the areas of plant tissue culture and molecular biology and their impact on plant improvement and biotechnology (Thorpe & Lorz, 1998). The titles of the plenary lectures were (1) "Plant Biotechnology Achievements and Opportunities at the Threshold of the 21st Century," (2) "Towards Sustainable Crops via International Cooperation," (3) "Signal Pathways in Plant Disease Resistance," (4) "Pharmaceutical Foodstuffs: Oral Immunization with Transgenic Plants," (5) "Plant Biotechnology and Gene Manipulation," and (6) "Use of Plant Roots for Environmental Remediation and Chemical Manufacturing". These titles not only clearly show where tissue culture was but where it was heading, as an equal partner with molecular biology as a tool in basic plant biology and in various areas of application. Later Congresses in Florida, USA (2002), Beijing, China (2006), and St Louis, Missouri, USA (2010) supported this view and clearly demonstrated the advances that were being made in these areas. Also see Datta (2007) and Stasolla and Thorpe (2011). As Schell (1995) pointed out, progress in applied plant biotechnology is fully matching and is in fact stimulating fundamental scientific progress.

REFERENCES

Ammirato, P. V. (1983). Embryogenesis. In D. A. Evans, W. R. Sharp, P. V. Ammirato, & Y. Yamada (Eds.), *Handbook of plant cell culture* (Vol. 1, pp. 82–123). New York: Macmillan.

Ball, E. (1946). Development in sterile culture of stems tips and subjacent regions of *Tropaeolum majus* L. and of *Lupinus albus* L. *American Journal of Botany, 33*, 301–318.

Beiderbeck, R., & Knoop, B. (1987). Two-phase culture. In F. Constabel, & I. K. Vasil (Eds.), *Cell culture and somatic cell genetics of plants* (Vol. 4, pp. 255–266). New York: Academic Press.

Bender, L., Kumar, A., & Neumann, K.-H. (1985). On the photosynthetic system and assimilate metabolism of *Daucus* and *Arachis* cell cultures. In K.-H. Neumann, W. Barz, & E. Reinhard (Eds.), *Primary and secondary metabolism of plant cell cultures* (pp. 24–42). Berlin: Springer-Verlag.

Bergmann, L. (1959). A new technique for isolating and cloning cells of higher plants. *Nature, 184*, 648–649.

Bhojwani, S. S., & Razdan, M. K. (1983). *Plant tissue culture: Theory and practice: Developments in crop science*. Amsterdam: Elsevier.

Binding, H. (1986). Regeneration from protoplasts. In I. K. Vasil (Ed.), *Cell culture and somatic cell genetics of plants* (Vol. 3, pp. 259–274). New York: Academic Press.

Bourgin, J. P., & Nitch, J. P. (1967). Obtention de *Nicotiana* haploides à partir de'étamines cultivées *in vitro*. *Annales de Physiologie Végétale, 9*, 377–382.

Braun, A. C. (1941). Development of secondary tumor and tumor strands in the crown-gall of sunflowers. *Phytopathology, 31*, 135–149.

Braun, A. C. (1950). Thermal inactivation studies on the tumor inducing principle in crown-gall. *Phytopathology, 40*, 3.

Braun, A. C., & White, P. R. (1943). Bacteriological sterility of tissues derived from secondary crown-gall tumors. *Phytopathology, 33*, 85–100.

Brodelius, P. (1985). The potential role of immobilisation in plant cell biotechnology. *Trends in Biotechnology, 3*, 280–285.

Brown, D. C. W., & Thorpe, T. A. (1986). Plant regeneration by organogenesis. In I. K. Vasil (Ed.), *Cell culture and somatic cell genetics of plants* (Vol. 3, pp. 49–65). New York: Academic Press.

Butcher, D. N. (1977). Plant tumor cells. In H. E. Street (Ed.), *Plant tissue and cell culture* (pp. 429–461). Oxford: Blackwell Scientific.

Carlson, P. S., Smith, H. H., & Dearing, R. D. (1972). Parasexual interspecific plant hybridization. *Proceedings of the National Academy of Sciences U.S.A., 69*, 2292–2294.

Chetrit, P., Mathieu, C., Vedel, F., Pelletier, G., & Primard, C. (1985). Mitochondrial DNA polymorphism induced by protoplast fusion in Cruciferae. *Theoretical and Applied Genetics, 69*, 361–366.

Cloutier, S., & Landry, B. S. (1994). Molecular markers applied to plant tissue culture. *In Vitro Cellular and Developmental Biology, 31P*, 32–39.

Cocking, E. C. (1960). A method for the isolation of plant protoplasts and vacuoles. *Nature, 187*, 927–929.

Collins, G. B., & Grosser, J. W. (1984). Culture of embryos. In I. K. Vasil (Ed.), *Cell culture and somatic cell genetics of plants* (Vol. 1, pp. 241–257). New York: Academic Press.

Conger, B. V. (Ed.), (1981). *Cloning agricultural plants via in vitro techniques*. Boca Raton, FL: CRC Press.

Constabel, F., & Vasil, I. K. (Eds.), (1987). *Cell culture and somatic cell genetics of plants*. (Vol. 4). New York: Academic Press.

Constabel, F., & Vasil, I. K. (Eds.), (1988). *Cell culture and somatic cell genetics of plants*. (Vol. 5). New York: Academic Press.

Crossway, A., Oakes, J. V., Irvine, J. M., Ward, B., Knauf, V. C., & Shewmaker, C. K. (1986). Integration of foreign DNA following microinjection of tobacco mesophyll protoplasts. *Molecular and General Genetics, 202*, 179–185.

D'Amato, F. (1978). Chromosome number variation in cultured cells and regenerated plants. In T. A. Thorpe (Ed.), *Frontiers of plant tissue culture 1978* (pp. 287–295). International Association of Plant Tissue Culture: Univ. of Calgary.

Datta, S. K. (2007). Impact of plant biotechnology in agriculture. In E. C. Pua, & M. R. Davey (Eds.), *Biotechnology in Agriculture and Forestry Transgenics crops IV* (Vol. 59, pp. 1–31). Berlin Heidelberg: Springer-Verlag.

Davey, M. R., Kumar, V., & Hammatt, N. (1994). *In vitro* culture of legumes. In I. K. Vasil, & T. A. Thorpe (Eds.), *Plant cell and tissue culture* (pp. 313–329). Dordrecht, The Netherlands: Kluwer.

Debergh, P. (1994). *In vitro* culture of ornamentals. In I. K. Vasil, & T. A. Thorpe (Eds.), *Plant cell and tissue culture* (pp. 561–573). Dordrecht, The Netherlands: Kluwer.

DeBlock, M., Herrera-Estrella, L., van Montague, M., Schell, J., & Zambryski, P. (1984). Expression of foreign genes in regenerated plants and in their progeny. *EMBO Journal, 3*, 1681–1689.

Deshayes, A., Herrera-Estrella, L., & Caboche, M. (1985). Liposome-mediated transformation of tobacco mesophyll protoplasts by an *Escherichia coli* plasmid. *EMBO Journal, 4,* 2731–2739.

Dix, P. J. (1994). Isolation and characterisation of mutant cell lines. In I. K. Vasil, & T. A. Thorpe (Eds.), *Plant cell and tissue culture* (pp. 119–138). Dordrecht, The Netherlands: Kluwer.

Dodds, J. (1989). Tissue culture for germplasm management and distribution. In J. I. Cohen (Ed.), *Strengthening collaboration in biotechnology: International agricultural research and the private sector* (pp. 109–128). Washington, D.C.: Bureau of Science and Technology, AID.

Dougall, D. K. (1987). Primary metabolism and its regulation. In C. E. Green, D. A. Somers, W. P. Hackett, & D. D. Biesboer (Eds.), *Plant tissue and cell culture* (pp. 97–117). New York: A. R. Liss.

Dudits, D., Györgyey, J., Bögre, L., & Bakó, L. (1995). Molecular biology of somatic embryogenesis. In T. A. Thorpe (Ed.), *In vitro embryogenesis in plants* (pp. 267–308). Dordrecht, The Netherlands: Kluwer.

Earle, E. D. (1978). Phytotoxin studies with plant cells and protoplasts. In T. A. Thorpe (Ed.), *Frontiers of plant tissue culture 1978* (pp. 363–372). Univ. of Calgary: International Association of Plant Tissue Culture.

Eilert, U. (1987). Methodology and aspects of application. In F. Constabel, & I. K. Vasil (Eds.), *Cell culture and somatic cell genetics of plants* (Vol. 4, pp. 153–196). New York: Academic Press. Elicitation.

Evans, D. A., Sharp, W. R., & Bravo, J. E. (1984). Cell culture methods for crop improvement. In W. R. Sharp, D. A. Evans, P. V. Ammirato, & Y. Yamada (Eds.), *Handbook of plant cell culture* (Vol. 2, pp. 47–68). New York: Macmillan.

Evans, D. A., Sharp, W. R., & Flick, C. E. (1981). Growth and behavior of cell cultures: Embryogenesis and organogenesis. In T. A. Thorpe (Ed.), *Plant tissue culture: Methods and applications in agriculture* (pp. 45–113). New York: Academic Press.

Fehér, A., & Dudits, D. (1994). Plant protoplasts for cell fusion and direct DNA uptake: Culture and regeneration systems. In I. K. Vasil, & T. A. Thorpe (Eds.), *Plant cell and tissue culture* (pp. 71–118). Dordrecht, The Netherlands: Kluwer.

Filner, P. (1978). Regulation of inorganic nitrogen and sulfur assimilation in cell suspension cultures. In T. A. Thorpe (Ed.), *Frontiers of plant tissue culture 1978* (pp. 437–442). Univ. of Calgary: International Association Plant Tissue Culture.

Flick, C. E. (1983). Isolation of mutants from cell culture. In D. A. Evans, W. R. Sharp, P. V. Ammirato, & Y. Yamada (Eds.), *Handbook of plant cell culture* (Vol. 1, pp. 393–441). New York: Macmillan.

Fowke, L. C. (1986). Ultrastructural cytology of cultured plant tissues, cells. In I. K. Vasil (Ed.), *Cell culture and somatic cell genetics of plants* (Vol. 3, pp. 323–342). New York: Academic Press.

Fowke, L. C. (1987). Investigations of cell structure using cultured cells and protoplasts. In C. E. Green, D. A. Somers, W. P. Hackett, & D. D. Biesboer (Eds.), *Plant tissue and cell culture* (pp. 17–31). New York: A. R. Liss.

Fowler, M. W. (1978). Regulation of carbohydrate metabolism in cell suspension cultures. In T. A. Thorpe (Ed.), *Frontiers of plant tissue culture 1978* (pp. 443–452). Univ. of Calgary: Intl. Assoc. Plant Tissue Culture.

Fowler, M. W. (1987). Process systems and approaches for large scale plant cell culture. In C. E. Green, D. A. Somers, W. P. Hackett, & D. D. Biesboer (Eds.), *Plant tissue and cell culture* (pp. 459–471). New York: A. R. Liss.

Fraley, R. (1992). Sustaining the food supply. *Bio/Technology, 10,* 40–43.

Fraley, R. T., Rogers, S. G., Horsch, R. B., Eichholtz, D. A., Flick, J. S., Fink, C. L., Hoffmann, N. L., & Sanders, P. R. (1985). The SEV system: A new disarmed Ti plasmid vector system for plant transformation. *Bio/Technology, 3,* 629–635.

Fujita, Y., & Tabata, M. (1987). Secondary metabolites from plant cells—pharmaceutical applications and progress in commercial production. In C. E. Green, D. A. Somers, W. P. Hackett, & D. D. Biesboer (Eds.), *Plant tissue and cell culture* (pp. 169–185). New York: A. R. Liss.

Fukuda, H. (1997). Xylogenesis: Initiation, progression, and cell death. *Annual Review of Plant Physiology and Plant Molecular Biology, 47,* 299–325.

Fukuda, H., & Komamine, A. (1985). Cytodifferentiation. In I. K. Vasil (Ed.), *Cell culture and somatic cell genetics of plants* (Vol. 2, pp. 149–212). New York: Academic Press.

Gamborg, O. L., Murashige, T., Thorpe, T. A., & Vasil, I. K. (1976). Plant tissue culture media. *In Vitro, 12,* 473–478.

Gaspar, T. (1995). The concept of cancer in *in vitro* plant cultures and the implication of habituation to hormones and hyperhydricity. *Plant Tissue and Culture Biotechnology, 1,* 126–136.

Gasser, C. S., & Fraley, R. T. (1989). Genetically engineering plants for crop improvement. *Science, 244,* 1293–1299.

Gautheret, R. J. (1934). Culture du tissus cambial. *C. R. Hebd. Seances Acad. Sc., 198,* 2195–2196.

Gautheret, R. J. (1935). *Recherches sur la culture des tissus végétaux.* Ph.D. Thesis, Paris.

Gautheret, R. J. (1939). Sur la possibilité de réaliser la culture indéfinie des tissus de tubercules de carotte. *C. R. Hebd. Seances Acad. Sc., 208,* 118–120.

Gautheret, R. J. (1942). Hétéro-auxines et cultures de tissus végétaux. *Bull. Soc. Chim. Biol., 24,* 13–41.

Gautheret, R. J. (1955). Sur la variabilité des propriétés physiologiques des cultures de tissus végétaux. *Rev. Gén. Bot., 62,* 5–112.

Gautheret, R. J. (1985). History of plant tissue and cell culture: A personal account. In I. K. Vasil (Ed.), *Cell culture and somatic cell genetics of plants* (Vol. 2, pp. 1–59). New York: Academic Press.

Grosser, J. W. (1994). *In vitro* culture of tropical fruits. In I. K. Vasil, & T. A. Thorpe (Eds.), *Plant cell and tissue culture* (pp. 475–496). Dordrecht, The Netherlands: Kluwer.

Guha, S., & Maheshwari, S. C. (1964). *In vitro* production of embryos from anthers of *Datura*. *Nature, 204,* 497.

Guha, S., & Maheshwari, S. C. (1966). Cell division and differentiation of embryos in the pollen grains of *Datura in vitro. Nature, 212,* 97–98.

Haberlandt, G. (1902). Kulturversuche mit isolierten Pflanzenzellen. *Sitzungsber. Akad. Wiss. Wien., Math.-Naturwiss. Kl., Abt., 1*(111), 69–92.

Hannig, E. (1904). Uber die Kultur von cruciferen embryonen ausserhalb den embryosacks. *Bot. Ztg., 62,* 45–80.

Harry, I. S., & Thorpe, T. A. (1994). *In vitro* culture of forest trees. In I. K. Vasil, & T. A. Thorpe (Eds.), *Plant cell and tissue culture* (pp. 539–560). Dordrecht, The Netherlands: Kluwer.

Heller, R. (1953). Recherches sur la nutrition minérale des tissus végétaux cultivé *in vitro. Annals Sci. Nat. Bot. Biol. Veg., 14,* 1–223.

Herzbeck, H., & Husemann, W. (1985). Photosynthetic carbon metabolism in photoautotrophic cell suspension cultures of *Chenopodium rubrum* L. In K.-H. Neumann, W. Barz, & E. Reinhard (Eds.), *Primary and secondary metabolism of plant cell culture* (pp. 15–23). Berlin: Springer-Verlag.

Hinchee, M. A.W., Corbin, D. R., Armstrong, C. L., Fry, J. E., Sato, S. S., Deboer, D. L., Petersen, W. L., Armstrong, T. A., Connor-Ward, D. V., Layton, J. G., & Horsch, R. B. (1994). Plant transformation. In I. K. Vasil, & T. A. Thorpe (Eds.), *Plant cell and tissue culture* (pp. 231–270). Dordrecht, The Netherlands: Kluwer.

Horsch, R. B., Fraley, R. T., Rogers, S. G., Sanders, P. R., Lloyd, A., & Hoffmann, N. (1984). Inheritance of functional foreign genes in plants. *Science, 223,* 496–498.

Horsch, R. B., Fry, J., Hoffman, N., Walroth, M., Eichholtz, D., Rogers, S., & Fraley, R. (1985). A simple and general method for transferring genes into plants. *Science, 227,* 1229–1231.

Hu, H., & Zeng, J. Z. (1984). Development of new varieties via anther culture. In P. V. Ammirato, D. A. Evans, W. R. Sharp, & Y. Yamada (Eds.), *Handbook of plant cell culture* (Vol. 3, pp. 65–90). New York: Macmillan.

Hughes, K. (1983). Selection for herbicide resistance. In D. A. Evans, W. R. Sharp, P. V. Ammirato, & Y. Yamada (Eds.), *Handbook of plant cell culture* (Vol. 1, pp. 442–460). New York: Macmillan.

Hüsemann, W. (1985). Photoautotrophic growth of cells in culture. In I. K. Vasil (Ed.), *Cell culture and somatic cell genetics of plants* (Vol. 2, pp. 213–252). New York: Academic Press.

Jacobs, M., Negruitiu, I., Dirks, R., & Cammaerts, D. (1987). Selection programmes for isolation and analysis of mutants in plant cell cultures. In C. E. Green, D. A. Somers, W. P. Hackett, & D. D. Biesboer (Eds.), *Plant tissue and cell culture* (pp. 243–264). New York: A. R. Liss.

Johri, B. M., & Bhojwani, S. S. (1965). Growth responses of mature endosperm in cultures. *Nature, 208,* 1345–1347.

Jones, M. G. K. (1994). *In vitro* culture of potato. In I. K. Vasil, & T. A. Thorpe (Eds.), *Plant cell and tissue culture* (pp. 363–378). Dordrecht, The Netherlands: Kluwer.

Jones, L. E., Hildebrandt, A. C., Riker, A. J., & Wu, J. H. (1960). Growth of somatic tobacco cells in microculture. *American Journal of Botany, 47,* 468–475.

Kaeppler, S. M., & Phillips, R. L. (1993). DNA methylation and tissue culture-induced variation in plants. *In Vitro Cellular Developmental Biology, 29P,* 125–130.

Kanta, K., Rangaswamy, N. S., & Maheshwari, P. (1962). Test-tube fertilization in flowering plants. *Nature, 194,* 1214–1217.

Karp, A. (1994). Origins, causes and uses of variation in plant tissue cultures. In I. K. Vasil, & T. A. Thorpe (Eds.), *Plant cell and tissue culture* (pp. 139–151). Dordrecht, The Netherlands: Kluwer.

Kartha, K. K. (1981). Meristem culture and cryopreservation methods and applications. In T. A. Thorpe (Ed.), *Plant tissue culture: Methods and applications in agriculture* (pp. 181–211). New York: Academic Press.

Kartha, K. K., & Engelmann, F. (1994). Cryopreservation and germplasm storage. In I. K. Vasil, & T. A. Thorpe (Eds.), *Plant cell and tissue culture* (pp. 195–230). Dordrecht, The Netherlands: Kluwer.

Kemp, H. A., & Morgan, M. R. A. (1987). Use of immunoassays in the detection of plant cell products. In F. Constabel, & I. K. Vasil (Eds.), *Cell culture and somatic cell genetics of plants* (Vol. 4, pp. 287–302). New York: Academic Press.

Klein, T. M., Wolf, E. D., Wu, R., & Sanford, J. C. (1987). High-velocity microprojectiles for delivering nucleic acids into living cells. *Nature, 327,* 70–73.

Kohlenbach, H. W. (1959). Streckungs- und Teilungswachstum isolierter Mesophyllzellen von *Macleaya cordata* (Wild.) R. Br. *Naturwissenschaften, 46,* 116–117.

Kohlenbach, H. W. (1966). Die Entwicklungspotenzen explantierter und isolierter Dauerzellen. I. Das Strechungs- und Teilungswachstum isolierter Mesophyllzellen von *Macleaya cordata, Z. Pflanzenphysiol., 55,* 142–157.

Komamine, A., Ito, M., & Kawahara, R. (1993). Cell culture systems as useful tools for investigation of developmental biology in higher plants: Analysis of mechanisms of the cell cycle and differentiation using plant cell cultures. In W. Y. Soh, J. R. Liu, & A. Komamine (Eds.), *Advances in developmental biology and biotechnology of higher plants* (pp. 289–310). *Proceedings* First Asia Pacific Conference on Plant Cell and Tissue Culture, Taedok Science Town, Taejon, Korea, 5–9 Sept. 1993. The Korean Society of Plant Tissue Culture.

Kong, L., Attree, S. M., Evans, D. E., Binarova, P., Yeung, E. C., & Fowke, L. C. (1998). Somatic embryogenesis in white spruce: Studies of embryo development and cell biology. In S. M. Jain, & P. K. Gupta (Eds.), *Somatic embryogenesis in woody plants* (Vol. 4, pp. 1–28). Dordrecht, The Netherlands: Kluwer.

Kotte, W. (1922). Kulturversuche mit isolierten Wurzelspitzen. *Beitr. Allg. Bot.*, *2*, 413–434.

Krikorian, A. D. (1994a). *In vitro* culture of root and tuber crops. In I. K. Vasil, & T. A. Thorpe (Eds.), *Plant cell and tissue culture* (pp. 379–411). Dordrecht, The Netherlands: Kluwer.

Krikorian, A. D. (1994b). *In vitro* culture of plantation crops. In I. K. Vasil, & T. A. Thorpe (Eds.), *Plant cell and tissue culture* (pp. 497–537). Dordrecht, The Netherlands: Kluwer.

Krikorian, A. D., & Berquam, D. L. (1969). Plant cell and tissue cultures: The role of Haberlandt. *Botanical Review*, *35*, 59–67.

Kutchin, T. M. (1998). Molecular genetics of plant alkaloid biosynthesis. In G. Cordell (Ed.), *The alkaloids* (Vol. 50, pp. 257–316). San Diego: Academic Press.

Kurz, W. G. W. (1988). Semicontinuous metabolite production through repeated elicitation of plant cell cultures: A novel process. In T. J. Mabry (Ed.), *Plant biotechnology* (pp. 93–103). Austin, TX: IC^2 Institute.

Laibach, F. (1929). Ectogenesis in plants: Methods and genetic possibilities of propagating embryos otherwise dying in the seed. *Journal of Heredity*, *20*, 201–208.

LaRue, C. D. (1936). The growth of plant embryos in culture. *Bulletin of the Torrey Botany Club*, *63*, 365–382.

LaRue, C. D. (1942). The rooting of flowers in culture. *Bulletin of the Torrey Botany Club*, *69*, 332–341.

LaRue, C. D. (1949). Culture of the endosperm of maize. *American Journal of Botany*, *36*, 798.

Larkin, P. J., & Scowcroft, W. R. (1981). Somaclonal variation—a novel source of variability from cell culture for plant improvement. *Theoretical and Applied Genetics*, *60*, 197–214.

Larkin, P. J., Brettell, R. I.S., Ryan, S. A., Davies, P. A., Pallotta, M. A., & Scowcroft, W. R. (1985). Somaclonal variation: Impact on plant biology and breeding strategies. In P. Day, M. Zaitlin, & A. Hollaender (Eds.), *Biotechnology in plant science* (pp. 83–100). New York: Academic Press.

Ledoux, L. (1965). Uptake of DNA by living cells. *Progress in Nucleic Acid Research and Molecular Biology*, *4*, 231–267.

Leonard, R. T., & Rayder, L. (1985). The use of protoplasts for studies on membrane transport in plants. In L. C. Fowke, & F. Constabel (Eds.), *Plant protoplasts* (pp. 105–118). Boca Raton, FL: CRC Press.

Letham, D. S. (1974). Regulators of cell division in plant tissues. The cytokinins of coconut milk. *Physiologia Planatrum*, *32*, 66–70.

Limasset, P., & Cornuet, P. (1949). Recherche du virus de la mosaïque du Tabac dans les méristèmes des plantes infectées. *C. R. Hebd. Seances Acad. Sci.*, *228*, 1971–1972.

Lindsey, K., & Yeoman, M. M. (1985). Dynamics of plant cell cultures. In I. K. Vasil (Ed.), *Cell culture and somatic cell genetics of plants* (Vol. 2, pp. 61–101). New York: Academic Press.

Loo, S. W. (1945). Cultivation of excised stem tips of *Asparagus in vitro*. *American Journal of Botany*, *32*, 13–17.

Miller, S. A., & Maxwell, D. P. (1983). Evaluation of disease resistance. In D. A. Evans, W. R. Sharp, P. V. Ammirato, & Y. Yamada (Eds.), *Handbook of plant cell culture* (Vol. 1, pp. 853–879). New York: Macmillan.

Miller, C., Skoog, F., Von Saltza, M. H., & Strong, F. M. (1955). Kinetin, a cell division factor from desoxyribonucleic acid. *Journal of the American Chemical Society*, *77*, 1392.

Monnier, M. (1995). Culture of zygotic embryos. In T. A. Thorpe (Ed.), *In vitro embryogenesis in plants* (pp. 117–153). Dordrecht, The Netherlands: Kluwer.

Morel, G. (1960). Producing virus-fee cymbidium. *American Orchid Society Bulletin, 29*, 495–497.

Morel, G., & Martin, C. (1952). Guérison de dahlias atteints d'une maladie á virus. *C. R. Hebd. Seances Acad. Sc., 235*, 1324–1325.

Muir, W. H., Hildebrandt, A. C., & Riker, A. J. (1954). Plant tissue cultures produced from single isolated plant cells. *Science, 119*, 877–878.

Muir, W. H., Hildebrandt, A. C., & Riker, A. J. (1958). The preparation, isolation and growth in culture of single cells from higher plants. *American Journal of Botany, 45*, 585–597.

Murashige, T. (1974). Plant propagation through tissue culture. *Annual Review of Plant Physiology, 25*, 135–166.

Murashige, T. (1978). The impact of plant tissue culture on agriculture. In T. A. Thorpe (Ed.), *Frontiers of plant tissue culture 1978* (pp. 15–26, 518–524). International Association of Plant Tissue Culture: Univ. of Calgary.

Murashige, T. (1979). Principles of rapid propagation. In K. W. Hughes, R. Henke, & M. Constantin (Eds.), *Propagation of higher plants through tissue culture: A bridge between research and application* (pp. 14–24). U.S. Dept. of Energy: Tech. Information Center.

Murashige, T. (1990). Practice with unrealized potential. In P. V. Ammirato, D. A. Evans, W. R. Sharp, & Y. P.S. Bajaj (Eds.), *Plant propagation by tissue culture Handbook of plant cell culture* (Vol. 5, pp. 3–9). New York: McGraw-Hill.

Murashige, T., & Skoog, F. (1962). A revised medium for rapid growth and bioassays with tobacco tissue cultures. *Physiologia Plantrum, 15*, 473–497.

Nagl, W., Pohl, J., & Radler, A. (1985). The DNA endoreduplication cycles. In J. A. Bryant, & D. Francis (Eds.), *The cell division cycle in plants* (pp. 217–232). Cambridge, UK: Cambridge Univ. Press.

Neumann, K.-H., Barz, W., & Reinhard, E. (Eds.), (1985). *Primary and secondary metabolism of plant cell cultures*. Berlin: Springer-Verlag.

Nitsch, J. P., & Nitsch, C. (1956). Auxin-dependent growth of excised *Helianthus tuberosus* tissues. *American Journal of Botany, 43*, 839–851.

Nobécourt, P. (1939a). Sur la pérennité et l'augmentation de volume des cultures de tissus végétaux. *C. R. Seances Soc. Biol. Ses Fil., 130*, 1270–1271.

Nobécourt, P. (1939b). Sur les radicelles naissant des cultures de tissus végétaux. *C. R. Seances Soc. Biol. Ses Fil., 130*, 1271–1272.

Nobécourt, P. (1955). Variations de la morphologie et de la structure de cultures de tissus végétaux. *Ber. Schweiz. Bot. Ges., 65*, 475–480.

Nomura, K., & Komamine, A. (1985). Identification and isolation of single cells that produce somatic embryos at a high frequency in a carrot suspension culture. *Plant Physiology, 79*, 988–991.

Nomura, K., & Komamine, A. (1995). Physiological and biochemical aspects of somatic embryogenesis. In T. A. Thorpe (Ed.), *In vitro embryogenesis in plants* (pp. 249–265). Dordrecht, The Netherlands: Kluwer.

Palmer, C. E., & Keller, W. A. (1994). *In vitro* culture of oilseeds. In I. K. Vasil, & T. A. Thorpe (Eds.), *Plant cell and tissue culture* (pp. 413–455). Dordrecht, The Netherlands: Kluwer.

Phillips, R. (1980). Cytodifferentiation. *International Review of Cytology, Supplement, 11A*, 55–70.

Potrykus, I., Shillito, R. D., Saul, M., & Paszkowski, J. (1985). Direct gene transfer: State of the art and future potential. *Plant Molecular Biology Report, 3*, 117–128.

Quack, F. (1961). Heat treatment and substances inhibiting virus multiplication in meristem culture to obtain virus-free plants. *Advances in Horticultural Science and their Application, Proceedings of the 15th International Horticulture Congress, 1958*(1), 144–148.

Raghavan, V. (1980). Embryo culture. *International Review of Cytology, Supplement, 11B*, 209–240.

Ranch, J. P., Rick, S., Brotherton, J. E., & Widholm, J. (1983). Expression of 5-methyltryptophan resistance in plants regenerated from resistant cell lines of *Datura innoxia*. *Plant Physiology, 71*, 136–140.

Rangan, T. S. (1982). Ovary, ovule and nucellus culture. In B. M. Johri (Ed.), *Experimental Embryology of Vascular Plants* (pp. 105–129). Berlin: Springer-Verlag.

Redenbaugh, K. (Ed.), (1993). *Synseeds: Applications of synthetic seeds to crop improvement.* Boca Raton, FL: CRC Press.

Reinert, J. (1958). Utersuchungen die Morphogenese an Gewebekulturen. *Ber. Dtsch. Bot. Ges., 71*, 15.

Reinert, J. (1959). Über die Kontrolle der Morphogenese und die Induktion von Adventivembryonen an Gewebekulturen aus Karotten. *Planta, 53*, 318–333.

Reynolds, J. F. (1994). *In vitro* culture of vegetable crops. In I. K. Vasil, & T. A. Thorpe (Eds.), *Plant cell and tissue culture* (pp. 331–362). Dordrecht, The Netherlands: Kluwer.

Robbins, W. J. (1922). Cultivation of excised root tips and stem tips under sterile conditions. *Botanical Gazette, 73*, 376–390.

Roberts, L. W. (1976). *Cytodifferentiation in plants: Xylogenesis as a model system.* Cambridge, UK: Cambridge Univ. Press.

Rottier, P. J. M. (1978). The biochemistry of virus multiplication in leaf cell protoplasts. In T. A. Thorpe (Ed.), *Frontiers of plant tissue culture 1978* (pp. 255–264). Univ. of Calgary: International Association of Plant Tissue Culture.

San, L. H., & Gelebart, P. (1986). Production of gynogenetic haploids. In I. K. Vasil (Ed.), *Cell culture and somatic cell genetics of plants* (Vol. 3, pp. 305–322). New York: Academic Press.

Schell, J. (1995). Progress in plant sciences is our best hope to achieve an economically rewarding, sustainable and environmentally stable agriculture. *Plant Tissue Culture Biotechnology, 1*, 10–12.

Schell, J. S. (1987). Transgenic plants as tools to study the molecular organization of plant genes. *Science, 237*, 1176–1183.

Schell, J., van Montague, M., Holsters, M., Hernalsteens, J. P., Dhaese, P., DeGreve, H., Leemans, J., Joos, H., Inze, D., Willmitzer, L., Otten, L., Wostemeyer, A., & Schroeder, J. (1982). *Plant cells transformed by modified Ti plasmids: A model system to study plant development. In Biochemistry of differentiation and morphogenesis*. Berlin: Springer-Verlag. 65–73.

Schieder, O., & Kohn, H. (1986). Protoplast fusion and generation of somatic hybrids. In I. K. Vasil (Ed.), *Cell culture and somatic cell genetics of plants* (Vol. 3, pp. 569–588). New York: Academic Press.

Schleiden, M. J. (1838). Beiträge zur Phytogenesis. *Müllers Arch. Anat. Physiol.*, 137–176.

Schwann, T. (1839). *Mikroscopishe Utersuchungen über die Übereinstimmung in der Struktur und dem Wachstum der Thiere und Pflanzen, No. 176.* Berlin: Oswalds.

Scowcroft, W. R., Brettell, R. I. S., Ryan, S. A., Davies, P. A., & Pallotta, M. A. (1987). Somaclonal variation and genomic flux. In C. E. Green, D. A. Somers, W. P. Hackett, & D. D. Biesboer (Eds.), *Plant tissue and cell culture* (pp. 275–286). New York: A. R. Liss.

Skoog, F., & Miller, C. O. (1957). Chemical regulation of growth and organ formation in plant tissue cultures *in vitro*. *Symposium of the Society of Experimental Biology, 11*, 118–131.

Skoog, F., & Tsui, C. (1948). Chemical control of growth and bud formation in tobacco stem segments and callus cultured *in vitro*. *American Journal of Botany, 35*, 782–787.

Stasolla, C., & Thorpe, T. A. (2011). Tissue culture; historical perspectives and applications. In A. Kumar, & S. K. Sopory (Eds.), *Applications of Plant Biotechnology* (in press). Dordecht, The Netherlands Kluwer Academic Publishers.

Steward, F. C., Mapes, M. O., & Mears, K. (1958). Growth and organized development of cultured cells. II. Organization in cultures grown from freely suspended cells. *American Journal of Botany, 45,* 705–708.

Stitt, M., & Sonnewald, U. (1995). Regulation of carbohydrate metabolism in transgenics. *Annual Review of Plant Physiology and Plant Molecular Biology, 46,* 341–368.

Street, H. E. (1969). Growth in organized and unorganized systems. In F. C. Steward (Ed.), *Plant physiology* (Vol. 5B, pp. 3–224). New York: Academic Press.

Street, H. E. (1977). Introduction. In H. E. Street (Ed.), *Plant tissue and cell culture* (pp. 1–10). Oxford, UK: Blackwell.

Suguira, M. (1997). *In vitro* transcription systems from suspension-cultured cells. *Annual Review of Plant Physiology and Plant Molecular Biology, 48,* 383–398.

Takebe, I., Labib, C., & Melchers, G. (1971). Regeneration of whole plants from isolated mesophyll protoplasts of tobacco. *Naturwissenschaften, 58,* 318–320.

Thompson, M. R., & Thorpe, T. A. (1990). Biochemical perspectives in tissue culture for crop improvement. In K. R. Khanna (Ed.), *Biochemical aspects of crop improvement* (pp. 327–358). Boca Raton, FL: CRC Press.

Thompson, M. R., & Thorpe, T. A. (1997). Analysis of protein patterns during shoot initiation in cultured *Pinus radiata* cotyledons. *Journal of Plant Physiology, 151,* 724–734.

Thorpe, T. A. (1980). Organogenesis *in vitro*: Structural, physiological, and biochemical aspects. *International Review of Cytology, 11A,* 71–111.

Thorpe, T. A. (1988). *In vitro somatic embryogenesis.* ISI Atlas of Science: Animal and Plant Science, 81–88.

Thorpe, T. A. (1990). The current status of plant tissue culture. In S. S. Bhojwani (Ed.), *Plant tissue culture: Applications and limitations* (pp. 1–33). Amsterdam: Elsevier.

Thorpe, T. A. (1993). Physiology and biochemistry of shoot bud formation *in vitro*. In W. Y. Soh, J. R. Liu, & A. Komamine (Eds.), *Advances in Developmental Biology and Biotechnology of Higher Plants* (pp. 210–224). The Korean Society of Plant Tissue Culture: *Proceedings* of the First Asia Pacific Conference on Plant Cells and Tissue Culture, Taedok Science Town, Taejon, Korea, 5–9 Sept. 1993.

Thorpe, T. A. (2007). History of plant tissue culture. *Molecular Biotechnology, 37,* 169–180.

Thorpe, T. A., & Lorz, H. (1998). *IAPTC Congress in retrospect: IXth International Congress on Plant Tissue and Cell Culture.* Jerusalem: Israel. June 14–19, 1998. Plant Tissue Culture Biotechnology 4, 121–124.

Tran Thanh Van, K. (1980). Control of morphogenesis by inherent and exogenously applied factors in thin cell layers. *International Review of Cytology, Supplement, 11A,* 175–194.

Tran Thanh Van, K., & Trinh, H. (1978). Morphogenesis in thin cell layers: Concept, methodology and results. In T. A. Thorpe (Ed.), *Frontiers of plant tissue culture 1978* (pp. 37–48). Univ. of Calgary: International Association of Plant Tissue Culture.

Trehin, C., Planchais, S., Glab, N., Perennes, C., Tregear, J., & Bergounioux, C. (1998). Cell cycle regulation by plant growth regulators: Involvement of auxin and cytokinin in the re-entry of Petunia protoplasts into the cell cycle. *Planta, 206,* 215–224.

Tukey, H. B. (1934). Artificial culture methods for isolated embryos of deciduous fruits. *Proceedings of the American Society of Horticultural Science, 32,* 313–322.

Tulecke, W. (1953). A tissue derived from the pollen of *Ginkgo biloba*. *Science, 117,* 599–600.

Tulecke, W., & Nickell, L. G. (1959). Production of large amounts of plant tissue by submerged culture. *Science, 130,* 863–864.

Uchimiya, H., Handa, T., & Brar, D. S. (1989). Transgenic plants. *Journal of Biotechnology, 12,* 1–20.

Van Overbeek, J., Conklin, M. E., & Blakeslee, A. F. (1941). Factors in coconut milk essential for growth and development of very young *Datura* embryos. *Science, 94,* 350–351.

Vasil, I. K. (Ed.), (1984). *Cell culture and somatic cell genetics of plants: Laboratory procedures and their applications.* (Vol. 1). New York: Academic Press.

Vasil, I. K., & Thorpe, T. A. (Eds.), (1994). *Plant cell and tissue culture.* Dordrecht, The Netherlands: Kluwer.

Vasil, I. K., & Vasil, V. (1994). *In vitro* culture of cereals and grasses. In I. K. Vasil, & T. A. Thorpe (Eds.), *Plant cell and tissue culture* (pp. 293–312). Dordrecht, The Netherlands: Kluwer.

Vasil, V., & Hildebrandt, A. C. (1965). Differentiation of tobacco plants from single, isolated cells in micro cultures. *Science, 150,* 889–892.

Verpoorte, R., van der Heijden, R., ten Hoopen, H. J.G., & Memclink, J. (1998). Metabolic engineering for the improvement of plant secondary metabolite production. *Plant Tissue Culture Biotechnology, 4,* 3–20.

Vöchting, H. (1878). *Über Organbildung im Pflanzenreich.* Bonn: Max Cohen.

White, P. R. (1934a). Potentially unlimited growth of excised tomato root tips in a liquid medium. *Plant Physiology, 9,* 585–600.

White, P. R. (1934b). Multiplication of the viruses of tobacco and Aucuba mosaics in growing excised tomato root tips. *Phytopathology, 24,* 1003–1011.

White, P. R. (1939a). Potentially unlimited growth of excised plant callus in an artificial nutrient. *American Journal of Botany, 26,* 59–64.

White, P. R. (1939b). Controlled differentiation in a plant tissue culture. *Bulletin of the Torrey Botany Club, 66,* 507–513.

White, P. R. (1963). *The cultivation of animal and plant cells* (2nd ed.). New York: Ronald Press.

Widholm, J. M. (1987). Selection of mutants which accumulate desirable secondary products. In F. Constabel, & I. K. Vasil (Eds.), *Cell culture and somatic cell genetics of plants* (Vol. 4, pp. 125–137). New York: Academic Press.

Wink, M. (1987). Physiology of the accumulation of secondary metabolites with special reference to alkaloids. In F. Constabel, & I. K. Vasil (Eds.), *Cell culture and somatic cell genetics of plants* (Vol. 4, pp. 17–42). New York: Academic Press.

Withers, L. A. (1985). Cryopreservation of cultured cells and meristems. In I. K. Vasil (Ed.), *Cell culture and somatic cell genetics of plants* (Vol. 2, pp. 253–316). New York: Academic Press.

Yamada, T., Shoji, T., & Sinoto, Y. (1963). Formation of calli and free cells in a tissue culture of *Tradescantia reflexa. Botany Magazine, 76,* 332–339.

Yamada, Y., Fumihiko, S., & Hagimori, M. (1978). Photoautotropism in green cultured cells. In T. A. Thorpe (Ed.), *Frontiers of plant tissue culture 1978* (pp. 453–462). Univ. of Calgary: International Association of Plant Tissue Culture.

Yeoman, M. M. (1987). Techniques, characteristics, properties, and commercial potential of immobilized plant cells. In F. Constabel, & I. K. Vasil (Eds.), *Cell culture and somatic cell genetics of plants* (Vol. 4, pp. 197–215). New York: Academic Press.

Yeoman, M. M., & Street, H. E. (1977). General cytology of cultured cells. In H. E. Street (Ed.), *Plant tissue and cell culture* (pp. 137–176). Oxford, UK: Blackwell.

Yeung, E. C., Thorpe, T. A., & Jensen, C. J. (1981). *In vitro* fertilization and embryo culture. In T. A. Thorpe (Ed.), *Plant tissue culture: Methods and applications in agriculture* (pp. 253–271). New York: Academic Press.

Zaenen, I., van Larebeke, N., Touchy, H., Van Montagu, M., & Schell, J. (1974). Super-coiled circular DNA in crown-gall inducing *Agrobacterium* strains. *Journal of Molecular Biology, 86,* 109–127.

Zenk, M. H. (1978). The impact of plant cell culture on industry. In T. A. Thorpe (Ed.), *Frontiers of plant tissue culture 1978* (pp. 1–13). Univ. of Calgary: International Association of Plant Tissue Culture.

Zenkteler, M. (1984). *In vitro* pollination and fertilization. In I. K. Vasil (Ed.), *Cell culture and somatic cell genetics of plants* (Vol. 1, pp. 269–275). New York: Academic Press.

Zenkteler, M., Misiura, E., & Guzowska, I. (1975). Studies on obtaining hybrid embryos in test tubes. In H. Y. Mohan Ram, J. J. Shaw, & C. K. Shaw (Eds.), *Form, structure and function in plants* (pp. 180–187). Meerut, India: Sarita Prakashan.

Zimmerman, R. H. (1986). Regeneration in woody ornamentals and fruit trees. In I. K. Vasil (Ed.), *Cell culture and somatic cell genetics of plants* (Vol. 3, pp. 243–258). New York: Academic Press.

Zimmerman, R. H., & Swartz, H. J. (1994). *In vitro* culture of temperate fruits. In I. K. Vasil, & T. A. Thorpe (Eds.), *Plant cell and tissue culture* (pp. 457–474). Dordrecht, The Netherlands: Kluwer.

Setup of a Tissue Culture Laboratory

The following describes some general considerations in the setup of a tissue culture laboratory at an academic institution where one is usually restricted to making the best use of existing laboratories. Kyte and Kleyn (1996) describe a commercial tissue culture laboratory design. Determining the location of the tissue culture laboratory is an important decision. Avoid locating it adjacent to laboratories that handle microorganisms or insects or facilities that are used to store seeds or other plant materials. Contamination from air vents and high foot traffic can be a problem. Foot traffic scuffs up the wax on floors as well as dust which help spread contaminants.

The tissue culture area should be kept clean at all times. This is important to ensure clean cultures and reproducible results. Avoid having potted plants in this area because they can be a source of mites and other contaminating organisms. Avoid field or greenhouse work immediately before entering the laboratory because mites and insects can be carried into the laboratory on hair and clothing. Personnel should shower and change clothes before entering the laboratory from the field or greenhouse.

In designing a laboratory for tissue culture use, arrange the work areas (media preparation/culture evaluation/record-keeping area, aseptic transfer area, and environmentally controlled culture area) so that there is a smooth traffic flow. The following is an outline of the major equipment and activity in each of the work areas of the laboratory.

Plant Tissue Culture. Third Edition. DOI: 10.1016/B978-0-12-415920-4.00002-5
23

WORK AREAS

Media Preparation/Culture Evaluation/Record-Keeping Area

1. Bench
2. Gas outlet
3. Hot plate and magnetic stirrer
4. Analytical and top-loading balances
5. pH meter
6. Refrigerator, freezer
7. Water purification and storage system
8. Dish-washing area
9. Storage facilities—glassware, chemicals
10. Autoclave (pressure-cooker will work for small media volume)
11. Low bench with inverted light and dissecting microscopes (avoid locating next to autoclaves or other high-humidity areas)
12. Fume hood
13. Desk and file cabinets
14. Desktop centrifuge, spectrophotometer, microwave (transformation studies and protoplast isolation)

Culture media may be conveniently prepared on a laboratory chemical bench with a pH meter, balances, and a sink in close proximity. The reagents and stock solutions should be located on shelves or in a refrigerator adjacent to the bench.

Glassware cleaning is a constant process because the turnover is usually very high. Culture tubes containing spent medium should be autoclaved at least 30 min, and the contents disposed of before washing. Autoclaved glassware should be promptly washed. Glassware should be scrubbed in warm, soapy water, rinsed three times with tap water, rinsed three times with distilled water, and placed in a clean area to dry. Generally dishwashers do not effectively clean culture vessels, and test tubes should be hand scrubbed (Table 2.1).

A low bench, table, file cabinet, and a desk are essential for culture evaluation and record-keeping. A desktop computer is very desirable for writing up reports.

Aseptic Transfer Area

1. Laminar air flow transfer hood and comfortable chair
2. Dissecting microscope
3. Gas outlet
4. Vacuum lines
5. Forceps, spatulas, scalpel, and disposable blades

A separate room for the transfer hood is ideal. This room should be designed so that there is positive-pressure air flow and good ventilation. It is also desirable to have a window to the outside or into the laboratory so that an individual spending long hours working in the hood may occasionally relieve eye strain.

TABLE 2.1 Glassware and Materials at Each Laboratory Station[a]

1 Bunsen burner, hose	4 Erlenmeyer flasks (250, 500 ml; 1, 2 liter)
1 tripod	6 Magenta boxes/baby food jars
1 pair autoclave gloves	8 slant racks
1 hot plate, magnetic stir bar	3 graduated cylinders (100, 500 ml; 1 liter)
1 parafilm roll	1 fingernail brush
1 spatula, large	1 sponge
1 spatula, double-prong	6 beakers (50, 100, 250, 600 ml; 1 liter)
1 book of matches	5 pipettes 91, 10 µl, 1; 5, 10 ml)
1 pipette filler	1 test tube rack, culture tubes (18 × 150 mm)
1 water squeeze bottle	6 latex gloves
10 sleeves Petri dishes	1 aluminum foil roll
3 volumetric flasks (100, 250, and 1000 ml)	

[a]A station will accommodate two students.

A laminar airflow hood can be expensive, and an alternative is to build one. One can improvise if a laminar air flow hood is not available and use a fume hood which has been thoroughly scrubbed down, sprayed with a hospital disinfectant, and has had the glass lowered to allow only enough room for the worker's hands and arms to do the transfers. When using a fume hood, do not turn it on as it will draw contaminated air from the room over the cultures and try to avoid having foot traffic in the area while transfers are under way. Cardboard boxes lined with aluminum foil or plastic containers can also function as clean, dead air spaces to do transfers. A HEPA filter can also be attached to a plastic storage box and used as a transfer hood. An article by Joe Kish in *Mushroom—The Journal of Wild Mushrooming* (Winter, 1997) describes the construction of a small hood (back issues are available for $5. Make checks to Maggie Rodgers at Fungal Cave Books, 1943 SE Locust, Portland, OR 97214; rogersm@aol.com). Complete plans including construction details for a 2- to 4-ft. hood can be found at http://envhort.ucdavis.edu/dwb/lamflohd.pdf.

Environmentally Controlled Culture Area

1. Shelves with lighting on a timer and controlled temperature
2. Incubators—with controlled temperature and light
3. Orbital shakers

High humidity in the culture room should be avoided because it increases contamination. Some culture rooms have dehumidifiers and air scrubbers.

Most cultures can be incubated in a temperature range of 25–27°C under a 16:8-h light:dark photoperiod controlled by clock timers. Experiments described in this manual use this as a standard culture condition. Illumination is from Gro-Lux or cool white 4-ft. long fluorescent lamps mounted 8 inches above the culture shelf and 12 inches apart. Light intensity varies depending on the age of the lights and whether the cultures are directly under them or off to one side. The light can be measured in foot-candles (fc; full sun is approximately 10,000 fc) or microeinsteins (μE) per second per square meter (1 μE $\cdot$ sec^{-1} $\cdot$ m^{-2} = 6.02 $\times$ 10^{17} photons^{-1} $\cdot$ m^{-2} = μmol $\cdot$ sec^{-1} $\cdot$ m^{-2}; full sun is approximately 2000 μE $\cdot$ sec^{-1} $\cdot$ m^{-2}). The range of light readings can be 40–200 fc or 20–100 μE $\cdot$ sec^{-1} $\cdot$ m^{-2}. A meter to measure foot-candles and a quantum radiometer–photometer light meter to measure microeinsteins per second per square meter may be used to measure the light level.

BIBLIOGRAPHY

Books

An extensive list of books on plant tissue culture and related topics is available from Agritech Consultants, Inc., P.O. Box 255, Shrub Oak, NY 10588, Fax/Phone: (914) 528–3469; e-mail: agricell@aol.com; HTTP://AgritechPublications.com/

Aitken-Christie, J., Kozai, T., & Smith, M. A. L. (1994). *Automation and environmental control in plant tissue cultures*. Boston: Kluwer.

Bajaj, Y. P. S. (Ed.), (1986). *Biotechnology in agriculture and forestry*. New York: Springer-Verlag. (42 different volumes on a range of topics in plant cell culture.)

Barz, W., Reinard, E., & Zenk, M. H. (Eds.), (1977). *Plant tissue culture and its biotechnological application*. New York: Springer-Verlag.

Bennett, A. B., & O'Neill, S. D. (Eds.), (1991). *Horticultural biotechnology*. New York: Wiley.

Bhojwani, S. S., & Razdan, M. K. (1983). *Plant tissue culture: Theory and practice*. New York: Elsevier.

Butcher, D. N., & Ingram, D. S. (Eds.), (1976). *Plant tissue culture*. Marietta, GA: Camelot.

Bonga, J. M., & VonAderkas, P. (1992). *In vitro culture of trees*. Boston: Kluwer.

Cassells, A. C., & Gahan, P. B. (2006). *Dictionary of plant tissue culture*. New York, London, Oxford: Food Products Press® (imprint of the Haworth Press, Inc.).

Celis, J. (1998). *Cell biology* (2nd ed.). New York: Academic Press.

Chupeau, Y., Caboche, M., & Henry, Y. (Eds.), (1989). *Androgenesis and haploid plants*. New York: Springer-Verlag.

Debergh, P. C., & Zimmerman, R. H. (Eds.), (1993). *Micropropagation technology and application*. Boston: Kluwer.

DeFossard, R. A. (Ed.), (1976). *Tissue culture for plant propagators*. Armidale, Australia: University of New England Printery.

Dhawan, V. (Ed.), (1989). *Applications of biotechnology in forestry and horticulture*. New York: Plenum.

Dixon, R. A. (1985). *Plant cell culture: A practical approach*. Oxford, England: IRL Press.

Dodds, J. H., & Roberts, L. W. (1985). *Experiments in plant tissue culture* (2nd ed.). New York: Cambridge Univ. Press.

Gamborg, O. L., & Wetter, L. R. (Eds.), (1975). *Plant tissue culture methods*. Ottawa: National Research Council of Canada.

Kyte, L., & Kleyn, J. (1996). *Plants from test tubes: An introduction to micropropagation* (3rd ed.). Portland, OR: Timber Press.

Laimer, M., & Rucker, W. (2003). *Plant tissue culture*. New York: Springer-Verlag.

Liang, G. H., & Skinner, D. Z. (Eds.), (2004). *Genetically modified crops, their development, uses, and risks*. New York, London, Oxford: Food Products Press® (imprint of the Haworth Press, Inc.).

Pierik, R. L. (2002). *In vitro culture of higher plants*. Dordrecht, The Netherlands: Kluwer Academic Publishers.

Pollard, J. W., & Walker, J. M. (Eds.), (1990). *Plant cell and tissue culture*. Clifton, NJ: Humana Press.

Razdan, M. K. (2002). *Introduction to plant tissue culture*. Enfield, NH: Science Publishers.

Reinert, J., & Bajaj, Y. P. S. (Eds.), (1977). *Applied and fundamental aspects of plant cell, tissue, and organ culture*. New York: Springer Verlag.

Reinert, J., & Yeoman, M. M. (Eds.), (1982). *Plant cell and tissue culture: A laboratory manual*. New York: Springer-Verlag.

Rubenstein, I., Gengenback, B., Phillips, R. L., & Green, L. E. (Eds.), (1980). *Genetic improvement of crops*. Minneapolis: Univ of Minnesota Press.

Stafford, A., & Warren, G. (Eds.), (1991). *Plant cell and tissue culture*. London: Open Univ. Press.

Street, H. E. (Ed.), (1973). *Plant tissue and cell culture*. Berkeley: Univ. of CA Press.

Street, H. E. (Ed.), (1974). *Tissue culture and plant science*. New York: Academic Press.

Thorpe, T. A. (Ed.), (1978). *Frontiers of plant tissue culture*. Calgary: Univ. of Calgary.

Thorpe, T. A. (2002). *In vitro embryogenesis in plants*. Dordrecht, The Netherlands: Kluwer Academic Publishers.

Thorpe, T. A. (Ed.), (1995). *In vitro embryogenesis in plants*. New York: Academic Press.

Torres, K. C. (1988). *Tissue culture techniques for horticulture*. New York: Van Nostrand Reinhold.

Trigiano, R. N., & Gray, D. J. (1996). *Concepts and laboratory exercises in tissue culture of vascular plants*. Boca Raton, FL: CRC Press.

Trigiano, R. N., & Gray, D. J. (2000). *Plant tissue culture concepts and laboratory exercises* (2nd ed.). Boca Raton, FL: CRC Press.

Vasil, I. K. (Ed.), (1980). *International review of cytology: Perspectives in plant cell and tissue culture (Suppl. 11, Part A)*. New York: Academic Press.

Vasil, I. K. (Ed.), (1980). *International review of cytology: Advances in plant cell and tissue culture (Suppl. 11, Part B)*. New York: Academic Press.

Wetherell, D. F. (1985). *Plant tissue culture*. Burlington, NC: Carolina Biology Readers No. 142, Carolina Biological Supply Co.

Journals

Acta Horticulturae
Agricell Report
American Journal of Botany
Biologia Plantarum
Biotechnology Letters
Biotechnology Advances
Biotechnology and Applied Biochemistry
Bio/Technology

BioTechniques
Botanical Gazette
Canadian Journal of Botany
Critical Reviews in Plant Sciences
Crop Science
Current Science
Electronic Journal of Biotechnology (http://www.ejb.org)
Engineering in Life Sciences
Genetics and Molecular Research
HortScience
In Vitro Cellular & Developmental Biology, Plant
Journal of Horticultural Science
Journal of Plant Physiology
Korean Journal Medicinal Crop Science
Molecular Breeding
Nature
Nature Biotechnology
Physiologia Plantarum
Plant Biology
Plant Biotechnology Reports
Plant and Cell Physiology
Plant Cell Reports
Plant Cell, Tissue and Organ Culture
Plant Growth Regulation
Plant Journal
Plant Molecular Biology
Plant Physiology
Plant Science
Plant Science Letters
Planta
Proceedings of the American Society for Horticultural Science
Proceedings of the International Plant Propagators Society
Propagation of Ornamental Plants
Protoplasma
Science
Scientia Horticulture
Southern Forests
The Plant Cell
Theoretical and Applied Genetics
Trends in Biotechnology
Trends in Plant Science
World Journal of Microbiology and Biotechnology

Listservers

An archive along with other information is available at the Plant-tc Listserve World Wide Web homepage <http://www.aagro.agri.umn.edu/plant-tc/listserv/>. A plant tissue culture information exchange homepage, http://aggiehorticulture.tamu.edu/tisscult/tcintro.html, is also a source of information.

Organizations

American Society for Horticulture Science
American Society of Agronomy
American Society of Plant Physiologists
International Association for Plant Tissue Culture and Biotechnology
International Society for Horticulture Science
Society for In Vitro Biology

Chapter 3

Media Components and Preparation

Chapter Outline

The selection or development of the culture medium is vital to success in tissue culture. No single medium will support the growth of all cells, and changes in the medium are often necessary for different types of growth response from a single explant. A literature search is useful for selecting the appropriate medium. Garcia *et al.* (2011) provide a useful guide on examining the effect of plant growth regulators, salt composition of the basal medium and a statistical analysis of the results. Likewise, Niedz and Evans (2007) can provide a guide for studying the effects of the MS inorganic salts on explant growth. If literature on the plant is not available, the development of a suitable medium is based on trial and error. The approach to developing the medium will depend on the purpose of the cell culture. Many of the media outlined in this manual can serve as useful starting points in developing a medium for a specific purpose, whether it is callus induction, somatic embryogenesis, anther culture, or shoot proliferation.

Appendixes I and II include useful measurement conversions and a review of solution preparation problems, respectively. A list of suppliers is in given Appendix III.

In general, the medium contains inorganic salts, and organic compounds like plant growth regulators, vitamins, a carbohydrate, hexitols, and a gelling agent.

Plant Tissue Culture. Third Edition. DOI: 10.1016/B978-0-12-415920-4.00003-7

31

In addition, the medium can also include amino acids, antibiotics, or natural complexes.

INORGANIC SALTS

The inorganic salt formulations can vary (Murashige, 1973; Huang & Murashige, 1976; Gamborg *et al.*, 1976; George *et al.*, 1987). Owen and Miller (1992) have carefully examined the widely used tissue culture media formulations and have pointed out minor errors in the original publications. Tables 3.1 and 3.2 outline the inorganic salt compositions of some of the commonly cited formulations. The Murashige and Skoog (MS) (1962) formulation is the most widely used (Smith & Gould, 1989) and will be the major salt formulation used in these exercises. The MS formulation was developed to insure that no increases in cell growth *in vitro* were due to the introduction of additional salts from plant tissue extracts which were being tested at that time. The MS formulation insured that the inorganic nutrients were not limiting to tobacco cell growth and organic supplements such as yeast extract, coconut milk, casein hydrolysate, and plant extracts were no longer essential sources of the inorganic salts. *The Science Citation Index* established the MS 1962 as a citation classic, as it has been extensively used in many publications on plant tissue culture. Very few articles in plant science can come close to this highly cited paper.

The distinguishing feature of the MS inorganic salts is their high content of nitrate, potassium, and ammonium in comparison to other salt formulations. Table 3.1 outlines the five MS inorganic salt stock solutions. These salt stocks are prepared at 100 times the final medium concentration, and each stock is added at the rate of 10 ml per 1000 ml of medium prepared. The NaFeEDTA stock should be protected from light by storing it in a bottle that is amber colored or wrapped in aluminum foil. Concentrated salt stocks enhance the accuracy and speed of media preparation.

Salt stocks are best stored in the refrigerator and are stable for several months. Always prepare stocks with glass-distilled or demineralized water and clearly label and date all stocks. Reagent-grade chemicals should always be used to ensure maximum purity. Several salts can be combined to minimize the number of stock solutions. The factors to consider in combining compounds are stability and coprecipitability. The nitrate stock will usually precipitate out and *must* be heated until the crystals are completely dissolved before using. Any stock that appears cloudy or has precipitates in the bottom should be discarded.

PLANT GROWTH REGULATORS

The type and concentration of plant growth regulators used will vary according to the cell culture purpose. A list of the most commonly used plant growth regulators, their abbreviations, and their molecular weights is provided in Table 3.3.

TABLE 3.1 Composition of the Five Inorganic Salt Stocks of the Murashige and Skoog Inorganic Formulation

Chemical	Concentration (g/liter stock)
Nitrate stock	
Ammonium nitrate (NH_4NO_3)	165.0
Potassium nitrate (KNO_3)	190.0
Sulfate stock	
Magnesium sulfate ($MgSO_4 \cdot 7H_2O$)	37.0
Manganous sulfate ($MnSO_4 \cdot H_2O$)	1.69
Zinc sulfate ($ZnSO_4 \cdot 7H_2O$)	0.86
Cupric sulfate ($CuSO_4 \cdot 5H_2O$)	0.0025
Halide stock	
Calcium chloride ($CaCl_2 \cdot 2H_2O$)	44.0
Potassium iodide (KI)	0.083
Cobalt chloride ($CoCl_233 \cdot 6H_2O$)	0.0025
PBMo stock	
Potassium phosphate (KH_2PO_4)	17.0
Boric acid (H_3BO_3)	0.620
Sodium molybdate ($Na_2MoO_4 \cdot 2H_2O$)	0.025
NaFeEDTA stock	
Ferrous sulfate ($FeSO_2 \cdot 7H_2O$)	2.784
Ethylenediamineteraacetic acid, disodium salt (Na_2EDTA)	3.724

An auxin (IAA, NAA, 2,4-D, or IBA) is required by most plant cells for division and root initiation. At high concentrations, auxin can suppress morphogenesis. The auxin 2,4-D is widely used for callus induction: IAA, IBA, and NAA are used for root induction.

Auxin stocks are usually prepared by weighing out 10 mg of auxin into a 200-ml beaker, adding several drops of 1 N NaOH or KOH until the crystals are dissolved (not more than 0.3 ml), rapidly adding 90 ml of double-distilled water, and increasing the volume to 100 ml in a volumetric flask (Huang & Murashige, 1976). Auxins can also be dissolved in 95% ethanol and diluted to volume; however, ethanol is toxic to plant tissues. The K-salts of auxin are more soluble in water (Posthumus, 1971).

TABLE 3.2 Inorganic Salt Formulation of Several Commonly Used Basal Salts for Plant Tissue Culture in Milligrams per Liter of Medium[a]

Chemical	White (1963)	B5[b]	N6[c]	WP[d]
NH_4NO_3				400
$(NH_4)2SO_4$		134	463	
$MgSO_4 \cdot 7H_2O$	720	246	185	370
KCl	65			
KNO_3	80	2528	2830	
KH_2PO_4			400	170
K_2SO_4				990
$NaH_2PO_4 \cdot H_2O$	19	150		
Na_2SO_4	200			
$CaCl_2 \cdot 2H_2O$		150	166	96
$Ca(NO_3)_2 \cdot 4H_2O$	300			556
$Na_2EDTA \cdot 2H_2O$		37.2	37.2	37.2
$FeSO_4 \cdot 7H_2O$		27.8	27.8	27.8
$Fe_2(SO_4)_3$	2.5			
H_3BO_3	1.5	3	1.6	6.2
$CoCl_2 \cdot 6H_2O$		0.025		
$CuSO_4 \cdot 5H_2O$	0.001	0.025		0.25
$MnSO_4 \cdot H_2O$		10		
$MnSO_4 \cdot 4H_2O$	7		4.4	22.3
MoO_3	0.0001			
$Na_2MoO_4 \cdot 2H_2O$		0.25		0.25
KI	0.75	0.75	0.8	
$ZnSO_4 \cdot 7H_2O$	3	2	1.5	8.6

[a]Owen and Miller (1992).
[b]B5, Gamborg et al. (1968).
[c]N6, Nitsch and Nitsch (1969).
[d]WP, Lloyd and McCown (1980).

TABLE 3.3 Common Plant Growth Regulators Used in Plant Tissue Culture

Plant growth regulator	Abbreviation	MW
Abscisic acid	ABA	264.3
Indole-3-acetic acid	IAA	175.2
Naphthaleneacetic acid	NAA	186.2
2,4-Dichlorophenoxyacetic acid	2,4-D	221.0
Indole-3-butyric acid	IBA	203.2
6-Furfurylaminopurine	Kinetin	215.2
6-Benzly-aminopurine	BA	225.2
N^6 (^{2}-isopentenyl)-adenine	2iP	203.3
Trans-6-(4-hydroxyl-3-methylbut-2-enyl) amino purine	Zeatin	219.2
Gibberellic acid	GA_3	346.4
Thidiazuron	TDZ	220.2
or 1-phenyl-3-(1,2,3-thiadiazol-5YL)-urea		

Make IAA stocks fresh weekly; IAA is degraded within a few days by light (Yamakawa *et al.*, 1979; Dunlap & Robacker, 1988) and within several hours to a few days by plant tissues (Epstein & Lavee, 1975). Auxins are thermostable at 110–120°C for up to 1 h (Posthumus, 1971; Yamakawa *et al.*, 1979). However, IAA is destroyed by low pH, light, oxygen, and peroxides (Posthumus, 1971); NAA and 2,4-D, which are synthetic auxins, are more stable than IAA, which is the naturally occurring auxin.

Cytokinins (kinetin, BA, zeatin, and 2iP) promote cell division, shoot proliferation, and shoot morphogenesis (Miller & Skoog, 1953; Miller, 1961). Thidiazuron (TDZ; N-phenyl-N^1-1,2,3-thiadiazol-5-ylurea) has cytokinin activity and is commercially used as a cotton defoliant. Thidiazuron has been effective in low concentrations to stimulate shoot formation (Sankhla *et al.*, 1996; Binzel *et al.*, 1996; Murthy *et al.*, 1998). Cytokinin stocks are prepared in a fashion similar to that for auxin stocks, except that 1 *N* HCl and a few drops of water are used to dissolve the crystals (Huang & Murashige, 1976). Gentle heating is usually required to completely dissolve crystals. Double-distilled water is rapidly added to avoid the crystals' falling out of solution. Bring the stock up to the desired volume in a volumetric flask. Cytokinin stocks can be stored for several months in the refrigerator. There can be some photochemical degradation in long-term experiments (Dekhuijzen, 1971).

Cytokinins (kinetin and zeatin) are thermostable; no breakdown products were detected after 1 h at 120°C (Dekhuijzen, 1971); 2iP and BA are stable for 20 min at 100°C.

Because it can inhibit callus growth and auxin-induced adventitious root formation, gibberellin (GA_3) is infrequently used in plant cell culture (Van Bragt & Pierik, 1971). However, it is useful in studies on morphogenesis. Stock solutions of GA_3 can be prepared by dissolving the crystals in water and adjusting the pH to 5.7. At an alkaline pH, GA is converted to an inactive isomer and in an acid pH and high temperature, GA_3 is also converted to biologically inactive forms (Van Bragt & Pierik, 1971). Solutions of GA_3 are not thermostable, and 20 min at 114°C reduces GA_3 activity by more than 90% (Van Bragt & Pierik, 1971). Stock solutions should be made up fresh before addition to the medium by filter sterilization.

Abscisic acid (ABA), a plant hormone involved in leaf and fruit abscission and dormancy, is useful in embryo culture. Abscisic acid is heat stable but light sensitive. The partial conversion of the *2-cis* isomer of ABA to the *2-trans* isomer of lower biological activity occurs in the light (Wilmar & Doornbos, 1971). Stock solutions can be prepared in water.

VITAMINS

Vitamins have catalytic functions in enzyme reactions. The vitamin considered important for plant cells is thiamine (B_1). Other vitamins, nicotinic acid (B_3) and pyridoxine (B_6), are added to cell culture media, as they may enhance cellular response. Vitamin stocks are best stored in a freezer and can be made up such that 10-ml aliquots are used per liter of medium prepared. *The vitamin stocks used in these exercises contain 5 mg of nicotinic acid and 5 mg pyridoxine-hydrochloride per 100 ml of water. The thiamine stock has 40 mg thiamine-hydrochloride in 1000 ml.* Other common vitamin formulations are those of White (1963, 1943) with in milligram-per-liter medium: 0.5 nicotinic acid, 0.1 pyridoxine-hydrochloride, and 0.1 thiamine-hydrochloride; B5 Gamborg (Gamborg *et al.*, 1976) with in milligram-per-liter medium: 100 inositol, 1.0 nicotinic acid, 1.0 pyridoxine-hydrochloride, and 10.0 thiamine-hydrochloride; Murashige and Skoog (1962) with in milligram-per-liter medium: 100 inositol, 0.5 nicotinic acid, 0.5 pyridoxine-hydrochloride, and 0.1 thiamine-hydrochloride. Most workers add vitamin stock solutions to the medium before autoclaving; however, for specific studies on vitamins, they should be filter sterilized (Ten Ham, 1971).

CARBOHYDRATES

Green cells in culture are generally not photosynthetically active and require a carbon source. Sucrose or glucose at 2–5% (w/v) is commonly used in cell culture. Other carbohydrate sources, such as fructose and starch, can also be used.

Lower levels of a carbohydrate may be used in protoplast culture, but much higher levels may be used for embryo or anther culture.

Sugars undergo caramelization if autoclaved too long (Peer, 1971; Ball, 1953) and will react with amino compounds (Maillard reaction). Caramelization occurs when sugars are heated, degrade, and form melanoidins, which are brown, high-molecular-weight compounds that can inhibit cell growth. A yellow to light brown color of an autoclaved medium is an indication that it was in the autoclave too long. The medium should be discarded.

HEXITOLS

The hexitol myo-inositol has been found to be important in tissue cultures (Pollard *et al.*, 1961; Steinhart *et al.*, 1962). Myo-inositol is an interesting hexitol involved in cyclitol biosynthesis, storage of polyhydric compounds as reserves, germination of seeds, sugar transport, mineral nutrition, carbohydrate metabolism, membrane structure, cell wall formation, hormonal homeostasis, and stress physiology (Loewus & Loewus, 1983). Myo-inositol is also considered as a growth enhancer *in vitro* and may be a carbohydrate source, but some feel it has vitaminlike action. Mannitol and sorbitol are hexitols, which are good osmotica for protoplast isolation.

GELLING AGENT

Many tissue culture experiments are conducted on some type of stationary support and a gelling agent is most commonly used. However, stationary supports can include filter paper, cotton, cheesecloth, vermiculite, and special membrane rafts with a liquid medium. The type of agar used to gell the medium can affect the response of your experiments (Griffis *et al.*, 1991; Debergh, 1983; Halquist *et al.*, 1983; Kacar *et al.*, 2010; Cassells & Collins, 2000). If the agar is unwashed or not purified, it will generally discolor the medium because it contains various impurities. Since agar is a product derived from seaweed, it can have physiological activity on the plant tissue. Sometimes dramatic differences in explant response can be observed by changing the brand of agar used. To minimize problems from agar impurities, purchase washed or purified agars. Gelrite is transparent in appearance and is a polysaccharide produced as a fermentation product from a *Pseudomonas* species (Kang *et al.*, 1982), and it is consistent in its composition. The exercises described in this manual use TC agar, Difco-Bacto agar, or Gelrite.

When melting agar over a hot plate or flame, keep the agar in motion either with a magnetic stir bar on the hot plate or by agitating the flask by hand. Use a heat-resistant glove on your hand because the flask can get very hot. The agar must be kept in motion or it will burn on the bottom of the flask. The agar must be completely dissolved before it is dispensed into the culture containers. The agar is dissolved when, after the Erlenmeyer flask is agitated, the medium that sheets off the interior glass surface does not have small granules of visible agar.

Remove the flask from the heat immediately because excessive heat after this point will cause the medium to boil out of the flask. Do not melt 1 liter of medium in a 1-liter Erlenmeyer flask; use a 2-liter Erlenmeyer flask to prevent media from boiling over. The medium is then dispensed in measured amounts in the culture container, which is capped and autoclaved. A dispensing burette can be used to accurately fill the culture container; however, because these are easily broken and are expensive, students should fill one container with the measured amount of water and use this as a guide to hand-fill the remaining containers. It helps to pour the hot medium from the Erlenmeyer flask into a 400- to 600-ml beaker before pouring it into the culture containers. Commercial cell culture laboratories use automatic media-dispensing equipment to rapidly fill culture containers.

The agar can also be melted in the autoclave in a foil-capped Erlenmeyer flask for 15 min at 121°C, 15 psi. When cool to the touch, the medium is dispensed aseptically into sterile containers within a transfer hood. If this method is used, the medium can be maintained in a water bath at 40°C to prevent it from solidifying before it is dispensed into the sterile containers.

When agar is not used, liquid media can be agitated on some type of a shaker. As mentioned earlier, explants can also be cultured on stationary liquid media usually on some type of a support like filter paper or membrane rafts. Preece (2010) discusses the use of stationary liquid medium for micropropagation. Interactions of the gelling agent concentration on the nutritional availability, hyperhydricity, and propagation rates are presented.

More recently bioreactor systems have been used to culture plant explants. The bioreactor is a sterile environment, and allows for exchange of the culture medium, as well as regulating air supply, pH and temperature. Bioreactor systems have been of high interest in commercial mass propagation of ornamentals to reduce labor costs (Hvoslef-Eide & Preil, 2004; Debnath, 2009; Fei & Weathers, 2011).

AMINO ACIDS

Amino acids and amines can be very important in morphogenesis. All L-forms of amino acids are the natural forms detected by the plant; L-tyrosine can contribute to shoot initiation (Skoog & Miller, 1957), L-arginine can facilitate rooting, and L-serine can be used in microspore cultures to obtain haploid embryos. Amides, such as L-glutamine and L-asparagine, sometimes significantly enhance somatic embryogenesis.

Casein hydrolysate, an enzymatic digest of milk protein (do not use the acid digest of milk proteins), was a common ingredient in many early media formulations as it provided a mixture of amino acids to enhance tissue response. Since enzymatic digests can result in slight differences in amino acid composition, addition of specific amino acids is preferred since it will more precisely define the ingredients in the medium. Today it is more common to examine individual amino acids for desirable tissue response.

ANTIBIOTICS

Because of excessive contamination problems with certain plant explants, many workers have incorporated fungicides and bactericides in the culture medium (Thurston *et al.*, 1979). Walsh (2003) examines antibiotics, and their action, origin, and antibiotic resistance. Generally, these additions have not been very useful because they can be toxic to the explant, and the contaminant can reappear as soon as the fungicides or bactericides are removed.

Transformation experiments using *Agrobacterium* make it necessary to incorporate antibiotics into the medium. Several antibiotics have been found not to be toxic to the explant and, at the same time, control or eliminate the *Agrobacterium*; commonly used antibiotics are timentin, carbenicillin (500 mg/liter), cefotaxime (300 µg/ml) and augmentin (250 mg/liter). The antibiotics are soluble in water, should be made up fresh, and should be added to the medium after autoclaving by filter sterilization.

NATURAL COMPLEXES

Many other additions to nutrient media serve various purposes. Antioxidants are sometimes used if there is excessive browning of the explant, and they retard oxidation of the explant. Examples of antioxidants are citric acid, ascorbic acid, pyrogallol, and phloroglucinol. Sometimes when there is excessive tissue discoloration of the medium and explant, absorbents are used. Two absorbents are polyvinylpyrrolidone (PVP) and activated charcoal (0.1–0.3%). Use activated, acid-neutralized charcoal.

Some good references on activated charcoal are Mohamed-Yasseen *et al.* (1995) and Wann *et al.* (1997). Recently Saenz *et al.* (2010) presented data on the effect of the source of activated charcoal on coconut embryogenic callus. Thomas (2008) suggests that activated charcoal may promote morphogenesis by absorbing inhibitory compounds and reducing toxic metabolites, phenolic exudation and accumulation. Activated charcoal releases substances naturally present in the charcoal that may promote growth. It also adsorbs vitamins and plant growth regulators from the medium, and may gradually release them to the explant. There may also be a beneficial effect related to darkening the medium.

A natural complex can be used when a defined medium fails to support a particular growth response. Natural complexes added to the medium generally make the medium undefined, since variations in growth-promoting or inhibitory compounds in these complexes are to be expected. Some examples and general concentrations of these natural complexes are coconut endosperm (CM, 10–20% v/v; Caplin & Steward, 1948), yeast extract (YE, 50–5000 mg/liter), malt extract (500 mg/liter), tomato juice (30%), orange juice (3–10%), banana (150 g/liter), potato extract (Chia-chun *et al.*, 1978), casein hydrolysate (30–3000 mg/liter, use enzyme digest), and fish emulsion (1 tsp/liter).

Much has been published on the compounds present in coconut milk, which is the liquid endosperm from *Cocos nucifera* L. A discussion in 1998 on the Plant-tc listserve (see also Yong *et al.*, 2009) evaluated the chemical composition and biological properties of coconut water including the sugars, vitamins, minerals, amino acids and phytohormones. Coconut water is the liquid from immature coconuts. As the coconut matures, the liquid turns to a jellylike substance called coconut milk. Coconut "water" is what is used as coconut "milk" in tissue culture. The coconut milk is boiled to precipitate the protein, cooled, filtered, and stored frozen until use. It is generally difficult to obtain coconut milk from coconuts in the grocery store, as they are mature and the endosperm is solid, or is real coconut milk. Coconut milk can be purchased from most chemical suppliers.

MEDIA pH

The pH of plant tissue culture media is generally adjusted to pH 5.5 to 6. Below 5.5, the agar will not gel properly and above 6.0, the gel may be too firm (Murashige, 1973). Media pH generally drops by 0.6 to 1.3 units after autoclaving (Sarma *et al.*, 1990). Cultures of some plant tissues cause a pH drop over time that is attributed to the production of organic acids or nitrogen utilization. Owen *et al.* (1991) examined media pH as influenced by the inorganic salts, carbohydrate source, gelling agent, activated charcoal, and medium storage method. All of these factors influenced the pH.

Adjust the medium pH with 1.0 or 0.1 N HC1 or NaOH by using a medicine dropper while keeping the medium stirred. Always adjust the pH before adding the agar.

MEDIUM PREPARATION

- Outline the medium to be prepared and check off the ingredients as they are added to the flask in which the medium is being prepared. Keep records on the date of preparation and use.
- In media preparation always use glass-distilled water—never tap water or tap-distilled water. Some water (~500 ml for final 1-liter volume) must always be in your flask before the stock solutions are added; otherwise, concentrated stocks will coreact and precipitate out.
- Never pour excess stocks back into the original stock solution container and never put excess sucrose or agar back into the original container. Always clean up spills around balance and work areas.
- Packaged powders of the MS salts and other media are available, eliminating the need to prepare stocks and measure ingredients. The suppliers of prepared plant tissue culture media are listed in Appendix III. Follow the manufacturer's directions for their use.

REFERENCES

Ball, E. (1953). Hydrolysis of sucrose by autoclaving media, a neglected aspect in the technique of culture of plant tissues. *Bulletin of the Torrey Botany Club, 80*, 409–411.

Binzel, M. L., Sankhla, N., Joshi, S., & Sankhla, D. (1996). Induction of direct somatic embryogenesis and plant regeneration in pepper (*Capsicum annuum* L.). *Plant Cell Reports, 15*, 536–540.

Caplin, S. M., & Steward, F. C. (1948). Effect of coconut milk on the growth of explants from carrot root. *Science, 108*, 655–657.

Cassells, A. C., & Collins, I. M. (2000). Characterization and comparison of agars and other gelling agents for plant tissue culture use. *Acta Hortic, 530*, 203–212.

Chia-chun, C., Tsun-wen, O., Hsu, C., Shu-min, C., & Chien-kang, C. (1978). A set of potato media for wheat anther culture. In *Proceedings of the symposium on plant and tissue culture* (pp. 51–55). Beijing: Science Press China (subsidiary of Van Nostrand-Reinhold, New York).

Debergh, P. C. (1983). Effect of agar brand and concentration on the tissue culture media. *Physiologia Plantarum, 59*, 270–276.

Debnath, S. C. (2009). Characteristics of strawberry plants propagated by in vitro bioreactor culture and ex vitro propagation method. *Engineering in Life Sciences, 9*(3), 239–264.

Debnath, S. C. (2010). A scaled-up system for in vitro multiplication of thidiazuron-induced red raspberry shoots using a bioreactor. *J. Hort. Sci & Biotech., 85*(2), 94–100.

Dekhuijzen, H. M. (1971). Sterilization of cytokinins. In J. Van Bragt, D. A. A. Mossel, R. L. M. Pierik, & H. Veldstra (Eds.), *Effects of sterilization on components in nutrient media* (pp. 129–132). Wageningen, The Netherlands: Kniphorst Scientific.

Dunlap, J. R., & Robacker, K. M. (1988). Nutrient salts promote light-induced degradation of indole-3-acetic acid in tissue culture media. *Plant Physiology, 88*, 379–382.

Epstein, E., & Lavee, S. (1975). Uptake and fate of IAA in apple callus tissue using IAA-1–14C. *Plant and Cell Physiology, 16*, 553–561.

Fei, L., & Weathers, P. J. (2011). From cells to field-ready plants: one-step micropropagation in a mist bioreactor. *In Vitro Cellular & Developmental Biology-Animal, 47*(1), S55–S55.

Gamborg, O. L., Miller, R. A., & Ojima, K. (1968). Nutrient requirements of suspension cultures of soybean root cells. *Experimental Cellular Research, 50*, 151–158.

Gamborg, O. L., Murashige, T., Thorpe, T. A., & Vasil, I. K. (1976). Plant tissue culture media. *In Vitro Cellular & Developmental Biology Plant, 12*, 473–478.

Garcia, R., Pacheco, G., Falcao, E., Borges, G., & Mansur, E. (2011). Influence of type of explant, plant growth regulators, salt composition of basal medium, and light on callogenesis and regeneration in *Passiflora suberosa* L. (Passifloraceae). *Plant Cell Tissue & Organ Culture, 106*(1), 47–54.

George, E. F., Puttock, D. J. M., & George, H. J. (1987). *Plant culture media: Formulations and uses.* (Vol. 1). Reading, England: Eastern Press.

Griffis, J. L., Wedekind, H., & Johnson, S. (1991). Effects of several commercially available solidifying agents on *in vitro* growth of *Alocasia bellota* "Alicia." *Proceedings of the Florida State Horticultural Society, 104*, 303–308.

Haluist, J. L., Hosier, M. A., & Read, P. E. (1983). A comparison of several gelling agents and concentrations on callus and organogenesis *in vitro*. *In Vitro, 19*, 248.

Hvoslef-Eide, T., & Preil, W. (2004). *Liquid culture systems for in vitro mass propagation of plants.* Dordrecht, The Netherlands: Springer.

Huang, L. C., & Murashige, T. (1976). Plant tissue culture media: Major constituents, their preparation and some applications. *TCA Manual, 3*, 539–548.

Kacar, Y. A., Bicen, B., & Varol, I. (2010). Gelling agents and culture vessels affect in vitro multiplication of banana plantlets. *Genetics & Molecular Research, 9*(1), 416–424.

Loewus, F. A., & Loewus, M. W. (1983). Myo-inositol: Its biosynthesis and metabolism. *Annual Review of Plant Physiology, 34*, 137–167.

Kang, K. S., Veeder, G. T., Mirrasoul, R. J., Kaneko, T., & Cottrell, I. W. (1982). Agar-like polysaccharide produced by a *Pseudomonas* species: Production and basic properties. *Applied Environmental Microbiology, 43*, 1086–1091.

Lloyd, C., & McCown, B. (1980). Commercially-feasible micropropagation of mountain laurel, *Kalmia latifilia*, by use of shoot-tip culture. *International Plant Propagators Society Proceedings, 30*, 421–427.

Miller, C. O. (1961). Kinetin and related compounds in plant growth. *Annual Review of Plant Physiology, 12*, 395–408.

Miller, C. O., & Skoog, F. (1953). Chemical control of bud formation in tobacco stem segments. *American Journal of Botany, 40*, 768–773.

Mohamed-Yasseen, Y., Barringer, S., Schloupt, R. M., & Splittstoesser, W. E. (1995). Activated charcoal in tissue culture: An overview. *Plant Growth Regulator Society of America, 23*(4), 206–213.

Murashige, T. (1973). Sample preparations of media. In P. F. Kruse, & M. K. Patterson (Eds.), *Tissue culture methods and applications* (pp. 698–703). New York: Academic Press.

Murashige, T., & Skoog, F. (1962). A revised medium for rapid growth and bioassays with tobacco tissue cultures. *Physiologia Plantarum, 15*, 473–497.

Murthy, B. N.S., Murch, S. J., & Saxena, P. (1998). Thidiazuron: A potent regulator of *in vitro* plant morphogenesis. *In Vitro Cellular and Developmental Biology-Plant, 34*, 267–275.

Niedz, R. P., & Evans, T. J. (2007). Regulating plant tissue growth by mineral nutrition. *In Vitro Cellular & Developmental Biology-Plant, 43*(4), 370–381.

Nitsch, J. P., & Nitsch, C. (1969). Haploid plants from pollen grains. *Science, 163*, 85–87.

Owen, H. R., & Miller, R. A. (1992). An examination and correction of plant tissue culture basal medium formulations. *Plant Cell Tissue & Organ Culture, 28*, 147–150.

Owen, H. R., Wengerd, D., & Miller, R. A. (1991). Culture medium pH is influenced by basal medium, carbohydrate source, gelling agent, activated charcoal, and medium storage method. *Plant Cell Reports, 10*, 583–586.

Preece, J. E. (2010). Micropropagation in stationary liquid media. *Propagation of Ornamental Plants, 10*(1), 183–187.

Peer, H. G. (1971). Degradation of sugars and their reactions with amino acids. In J. Van Bragt, D. A. A. Mossel, R. L. M. Pierik, & H. Veldstra (Eds.), *Effects of sterilization on components in nutrient media* (pp. 105–113). Wageningen, The Netherlands: Kniphorst Scientific.

Pollard, J. K., Santz, E. M., & Steward, F. C. (1961). Hexitols in coconut milk: Their role in nurture of dividing cells. *Plant Physiology, 36*, 492–501.

Posthumus, A. C. (1971). Auxins. In J. Van Bragt, D. A. A. Mossel, R. L. M. Pierik, & H. Veldstra (Eds.), *Effects of sterilization on components in nutrient media* (pp. 125–128). Wageningen, The Netherlands: Kniphorst Scientific.

Saenz, L., Herrera-Herrera, G., Uicab-Ballote, F., Chan, J. L., & Oropeza, C. (2010). Influence of form of activated charcoal on embryogenic callus formation in coconut (*Cocos nucifera* L.). *Plant Cell Tissue & Organ Culture, 100*(3), 301–308.

Sankhla, D., Davis, T. D., & Sankhla, N. (1996). *In vitro* regeneration of silktree (*Albizzia julibrissin*) from excised roots. *Plant Cell Tissue & Organ Culture, 44*, 83–86.

Sarma, K. S., Maesato, K., Hara, T., & Sonoda, Y. (1990). Effect of method of agar addition on post-autoclave pH of the tissue culture media. *Annals of Botany, 65*, 37–40.

Skoog, F., & Miller, C. O. (1957). Chemical regulation of growth and organ formation in plant tissues cultured *in vitro*. *Symposium of the Society of Experimental Biology, 11*, 118–131.

Smith, R. H., & Gould, J. H. (1989). Introductory essay. In J. Janick (Ed.), *Classic papers in horticultural science* (pp. 52–90). Englewood Cliffs, NJ: Prentice-Hall.

Steinhart, C., Anderson, L., & Skoog, F. (1962). Growth promoting effect of cyclitols on spruce tissue cultures. *Plant Physiology, 37*, 60–66.

Ten Ham, E. J. (1971). Vitamins. In J. Van Bragt, D. A. A. Mossel, R. L. M. Pierik, & H. Veldstra (Eds.), *Effects of sterilization on components in nutrient media* (pp. 121–123). Wageningen, The Netherlands: Kniphorst Scientific Bookshop.

Thomas, T. D. (2008). The role of activated charcoal in plant tissue culture. *Biotechnology Advances, 26*(6), 618–631.

Thurston, K. C., Spencer, S. J., & Arditti, J. (1979). Phytotoxicity of fungicides and bactericides in orchid culture media. *American Journal of Botany, 66*, 825–835.

Van Bragt, J., & Pierik, R. L. M. (1971). The effect of autoclaving on the gibberellin A₃. In J. Van Bragt, D. A. A. Mossel, R. L. M. Pierik, & H. Veldstra (Eds.), *Effects of sterilization on components in nutrient media* (pp. 133–137). Wageningen, The Netherlands: Kniphorst Scientific.

Walsh, L. (2003). *Antibiotics: Actions, origins, resistance*. Herndon, VA: ASM Press.

Wann, S. R., Veazey, R. L., & Kaphammer, J. (1997). Activated charcoal does not catalyse sucrose hydrolysis in tissue culture media. *Plant Cell Tissue and Organ Culture, 50*, 221–224.

White, P. R. (1943). *A handbook of plant tissue culture*. Lancaster, PA: Jaques Cattell Press.

White, P. R. (1963). *The cultivation of animal and plant cells* (2nd ed.). New York: Ronald Press.

Wilmar, J. C., & Doornbos, T. (1971). Stability of abscisic acid isomers to heat sterilization and light. In J. Van Bragt, D. A. A. Mossel, R. L. M. Pierik, & H. Veldstra (Eds.), *Effects of sterilization on components in nutrient media* (pp. 139–147). Wageningen, The Netherlands: Kniphorst Scientific.

Yamakawa, T., Kurahashi, O., Ishida, K., Kato, S., Kodama, T., & Minoda, Y. (1979). Stability of indole-3-acetic acid to autoclaving, aeration and light illumination. *Agriculture & Biochemical Journal, 43*(4), 879–880.

Yong, J. W. H., Ge, Liya, Fei Ng, Yan, & Tan, Swee Ngin (2009). The chemical composition and biological properties of coconut (*Cocos nucifera* L.) water. *Molecules, 14*(12), 5144–5164.

Explant Preparation

The explant is a piece of plant tissue placed into tissue culture. An explant can develop a callus as a wound response that consists of unorganized, dividing cells. Additionally, callus can be produced without wounding by germinating some seeds on a medium containing a plant growth regulator like 2,4-D. Callus cells vary in size, shape, pigmentation, and sometimes in genetic expression. These cells are highly differentiated in that they have a large central vacuole and the nucleus is to the side. This is in contrast to undifferentiated, meristematic cells that are isodiametric, small, lack a prominent vacuole, are cytoplasmic, and have a large central nucleus. These meristematic cells are sometimes initiated in callus masses and are referred to as meristemoid regions. Meristemoids can give rise to adventitious roots, shoots, or somatic embryos.

Factors in explant selection include consideration of the following:

1. Physiological or ontogenic age of the organ that is to serve as the explant source
2. Season in which the explant is obtained
3. Size and location of the explant
4. Quality of the source plant
5. Ultimate goal of cell culture
6. Plant genotype

Plant Tissue Culture. Third Edition. DOI: 10.1016/B978-0-12-415920-4.00004-9
45

EXPLANT AGE

The age of the explant can be very important, as physiologically younger tissue is generally more responsive *in vitro*. In many cases, older tissue will not form callus that is capable of regeneration. In addition, younger tissue is usually the newest formed and is generally easier to surface disinfect and establish clean cultures.

SEASON

The season of the year can have effects on contamination and response in culture. For example, buds or shoots taken during the spring of the year when the shoots are in a flush state of growth are more responsive than dormant buds. As the season of the year passes from spring, summer, and fall to winter the explant generally does not respond as well in culture. Tissue that is physiologically dormant is generally unresponsive in culture until the dormancy requirement is met; sometimes the dormancy response can be met in culture such as chilling bulb scale explants or embryos to break dormancy. Additionally, contamination rates also increase as the summer progresses; fall and winter contamination can increase to 100%.

EXPLANT SIZE

The explant size has an effect on the response of the tissue. Generally, the smaller the explant, the harder it is to culture. The culture medium usually has to have additional components. The larger explants probably contain more nutrient reserves and plant growth regulators to sustain the culture. Tran Thah Van (1977) published an interesting report on thin epidermal slices of tobacco stem tissue and varying morphogenetic potential depending on whether the explant is taken from the base, middle, or top of the stem. Plants have different hormonal balances throughout the plant and depending on the location of the explant, the explant can have a different endogenous level of plant growth regulators. Internal differences in hormone balance in the tissue can result in varying *in vitro* responses.

PLANT QUALITY

It is advisable to obtain explants from plants which are healthy as compared to plants under nutritional or water stress or plants which are exhibiting disease symptoms. In some instances such as when establishing virus-free plants, the plant from which the explant is harvested has a virus or multiple viruses.

GOAL

Depending on what type of a response is desired from the cell culture, the choice of explant tissue will vary. Any piece of plant tissue can be used as an explant (Fig. 4.1). For example, if clonal propagation is the goal, then the explant will

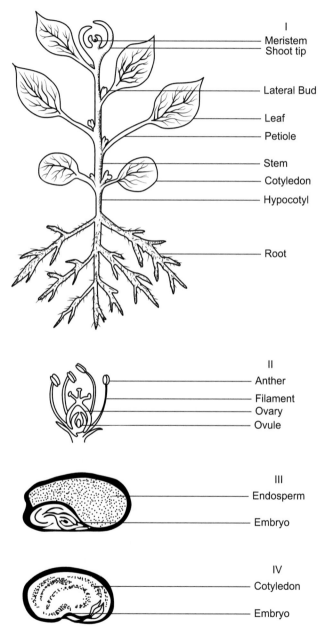

FIGURE 4.1   Schematic drawings (from top to bottom) of a plant, a flower, and monocotylede-
nous and dicotyldenous seeds indicate potential explant tissues.

usually be a lateral or terminal bud or shoot. For callus induction, pieces of the cotyledon, hypocotyl, stem, leaf, or embryo are usually used. Excellent explants for callus induction are seedling tissues from aseptically germinated seeds or immature inflorescences. Leaf tissue from the aseptically germinated seed is a good source of tissue for protoplast isolation. To produce haploid plants or callus, the anther or pollen is cultured. The exercises in this text will use a variety of explants.

GENOTYPE

Among any plant genus, there are usually large differences in the genotypes, cultivars, or species and their response in cell culture. Some genotypes are not responsive in culture, or recalcitrant, while others easily respond to produce callus or shoots. Testing numerous genotypes of a crop or ornamental species is generally a major experimental parameter to identify those that will respond in culture.

ASEPTIC TECHNIQUE

Explants require surface-disinfestation before they can be placed in culture on the nutrient agar. This is generally accomplished by using diluted commercial chlorine bleach. Some explants, such as very small seeds or fern spores, are surface-disinfected in conical, capped centrifuge tubes and require centrifugation to pellet the seeds and decanting off the solutions with a pipette. Explants that float in the disinfestant can be wrapped in squares of cheesecloth to prevent their floating out of the beaker or test tube as the solutions are changed.

A general procedure for preparing the explant is as follows:

1. Wash the explant in warm, soapy water and rinse in tap water. This procedure is very beneficial for stem, leaf, and shoot tip explants from the field or greenhouse because it removes surface contaminants.
2. Sometimes a brief alcohol rinse or swabbing with alcohol-wetted cheesecloth is appropriate especially with surfaces that are hairy or coated with thick wax.
3. Immerse the explant in the chlorine bleach solution, which should always be made up fresh. Always add 1–2 drops of Tween-20, detergent, or other wetting agent per 100 ml of bleach solution. A 10% bleach solution is prepared by adding 10 ml of chlorine bleach to a graduated cylinder and diluting with water to 100 ml. When disinfesting in culture tubes, pour the bleach solution into the culture tube cap and then into the culture tube; this helps to disinfest the cap. Place your hand over the tube and mix. Cap and agitate occasionally for 5–30 min for disinfestation. Commercial chlorine bleach once opened can age and lose its effectiveness. For this reason small containers of commercial bleach are preferable to large ones.
4. Decant the bleach solution and rinse the explant in sterile water three to five times. This step is carried out in a transfer hood.

Some tissues are more difficult than others for establishing clean cultures. The concentration of chlorine bleach and the length of time the tissue is in the bleach can vary. Delicate, succulent tissue may be clean after 10 min in a 10% solution. Fern spores may only require a 3–5% bleach for 3–4 min. Some seeds may require 30–50% bleach for 20–60 min. Any cut explant such as a stem or leaf that is surface sterilized will almost always shows tissue damage from the surface sterilization. These damaged pieces should be removed before culture.

Contamination that results from improperly sterilized tissue will generally arise from the explant and be located in the medium adjacent to the explant. Contamination due to poor technique generally will appear over the entire agar surface. Contamination of cultures by fungi appears as a fuzzy growth. Bacterial contamination appears as smooth pink, white, or yellow colonies. If contamination is due to poor technique, contaminated transfer hood filters, or improperly sterilized media it will be scattered on or in the medium. Contamination from insects will generally appear as tracks across the medium, which are visible due to bacterial or fungal growth on the insect tracks. For a more detailed discussion on contamination *in vitro*, refer to Chapter 5.

ASEPTIC GERMINATION OF SEEDS

Purpose: To gain experience in media preparation, disinfestation of explants, aseptic culture techniques, and establish cultures for callus induction experiments.

Carrot, Sunflower, Broccoli, Cotton, or Any Seed of Choice

Medium Preparation: 1-liter equivalent, Seed Germination Medium.

1. Into a 2000-ml Erlenmeyer flask pour 500 ml of deionized, distilled water.
2. Mix in 10 ml each Murashige and Skoog salts: nitrates, halides, NaFeEDTA, sulfates, and PBMo.
3. Adjust volume to 1000 ml. Adjust pH to 5.7 using 1 N HCl or NaOH.
4. Add 8 g/liter TC agar or agar brand of choice. Melt.
5. Distribute 10 ml/culture tube (25 × 150 mm). Cap.
6. Prepare sterile rinse water: 250-ml Erlenmeyer flask with ~100 ml distilled water. Cap.
7. Autoclave for 15 min at 121°C, 15 psi.

Explant Preparation
Carrot, sunflower, and broccoli

1. Place approximately 10 seeds on cheesecloth (~8 × 8 cm). Twist the cloth around the seeds and place in a 25 × 150-mm culture tube.
2. Rinse 2–3 min in 70% alcohol. Decant alcohol.

3. Add 20% chlorine bleach solution (20 ml chlorine bleach + 80 ml water + 2 drops Tween-20) to the culture tube. Cap.
4. Agitate for 15 min. The bleach may spot your clothes, so be careful.
5. Decant the bleach solution and aseptically rinse the seed three times in sterile water using a transfer hood.
6. Remove the seeds from the culture tube, place in a sterile Petri dish, unwrap, and place on medium. A sterile spatula is useful for planting small carrot seeds, and forceps can be used for larger seeds. Wrap the junction of the cap and the test tube with Parafilm. Place the cultures on the culture shelf for growth.

The sunflower seeds will germinate within a week, but carrot seeds may take 2 weeks. All parts of the seedling can be used for callus induction in 1–3 weeks.

Cotton

1. Wrap seeds in cheesecloth.
2. Add 20% chlorine bleach solution (20 ml chlorine bleach + 80 ml water + 2 drops Tween-20). Cap.
3. Sterilize for 20 min.
4. At the end of 20 min, decant the bleach solution and rinse the seeds in sterile water three times.
5. Unwrap the seeds in a Petri dish and plant them in culture tubes. Cotton seeds are dipped in 95% ethanol and flamed before planting.
6. Incubate the cultures on the culture shelf.

Some cultivars of cotton are difficult to disinfect; however, if these seeds are placed in sulfuric acid for 15 min, rinsed by pouring in an acid-resistant funnel, scrubbed in warm soapy water, placed in 50% chlorine bleach for 40 to 60 min while agitating, and rinsed three times with sterile water, even the most difficult seed batches can be cleaned.

Douglas Fir

Medium Preparation: sterile vermiculite

1. Pour a capful of fine vermiculite into each of 10 test tubes (25 × 150 mm).
2. Moisten with 10–15 ml deionized, distilled water. Cap.
3. Autoclave for 15 min at 121°C, 15 psi.

Explant Preparation

1. Wrap seeds in cheesecloth.
2. Place in a test tube.
3. Cover with a 20% chlorine bleach solution (20 ml chlorine bleach + 80 ml water + 2 drops Tween-20). Cap.
4. Sterilize for 15 min.
5. In the hood, pour off the bleach solution and rinse three times in sterile water.

6. Remove the cheesecloth in a Petri dish and plant the seeds.
7. Incubate the cultures on the culture shelf.

Plants will be ready in 2–4 weeks to use in the "Primary Morphogenesis: Douglas Fir" exercise.

Observations

The scientific names for the plants used in this exercise are *Daucus carota* L. (carrot), *Gossypium hirsutum* L. (cotton), *Pseudotsuga menziesii* Mirb. Franco (Douglas Fir), and *Helianthus annuus* L. (sunflower). There is also a specific name for the cultivar or variety you are using. It is important that it be recorded. Suggested cultivars are carrot, Danvers Red Core; cotton, Coker 312; and sunflower, Triumph 560. Different cultivars often respond differently when cultured. Generally, there is little to no contamination, and 70–80% of the seeds should germinate.

After 1 week note any contamination. If contamination is observed, try to ascertain whether it arose from the explant or from poor technique in planting. The sterile seedlings will be used as explants in the callus induction and explant orientation exercises.

BIBLIOGRAPHY

Beck, C. (2010). *An introduction to plant structure and development* (2nd ed.). Boston: Cambridge University Press.

Bell, A. D. (2008). *Plant form: an illustrated guide to flowering plant morphology*. Portland, London: Timber Press.

Cronquist, A. (1973). *Basic botany*. New York: Harper & Row.

Esau, K. (1988). *Plant anatomy* (3rd ed.). New York: Wiley.

Fahn, A. (1990). *Plant anatomy* (4th ed.). New York: Pergamon.

Tran Thanh Van, K. (1977). Regulation of morphogenesis. In W. Barz, E. Reinhard, & M. H. Zenk (Eds.), *Plant tissue culture and its biotechnological application* (pp. 367–385). New York: Springer-Verlag.

Weier, T. F., Stocking, C. R., & Barbour, M. G. (1970). *Botany: An introduction to plant biology* (4th ed.). New York: Wiley.

Contamination

The establishment of sterile cultures can be a major challenge with some plant materials. Additionally, the initial process of cleaning and disinfecting the plant material, especially if the parent plant is rare and the supply is limited, can be time-consuming and expensive. Not only can the establishment of clean cultures be a problem, but also subsequent loss of cultures due to insect infestations, which bring in fungal and bacterial contamination, can be devastating.

Sources of microbial contamination can include:

1. Explant
2. Personnel
3. Laminar air flow hood
4. Mites, ants, and other insects
5. Media
6. Instruments
7. Room air handling system

EXPLANT

The explant, or piece of plant tissue to be cultured, is often the major source of contaminants. There are microorganisms on the surface, in small crevices, and

Plant Tissue Culture. Third Edition. DOI: 10.1016/B978-0-12-415920-4.00005-0
53

between the outer layers of bulb scales, developing leaves of buds, and so on. Surfaces covered with thick wax such as the succulents and epidermal hairs (trichomes) can trap microorganisms. Additionally, many plants have microbial contamination within the vascular system, some seeds (Donnarumma *et al.*, 2011), and intercellular spaces in leaf mesophyll, xylem vessel lumen and intercellular spaces of callus (Miyazaki *et al.*, 2011).

To remove surface contaminants, prewash the explant in warm, soapy water to remove soil and dust and to enhance the contact of the disinfectant. To enhance disinfestation, use a wetting agent such as Tween-20 or a detergent like dishwashing soap, which acts as a surfactant. Usually 1–2 drops of the surfactant per 100 ml of the disinfecting agent are adequate. To enhance contact of the disinfesting agent with the explant, shake, stir, or agitate the explant while it is being disinfested. It has been suggested (listserve plant-tc discussions) that sonication treatment during disinfestation can enhance disinfestation of the explant. The sonication time can vary from 2 min for soft tissue to 20 min for seeds. If the tissue is covered with epidermal hairs, which can trap air bubbles, evacuate under a vacuum. The disinfesting agent should be rinsed from the explant with 3–5 rinses in sterile water.

Trial and error or a literature reference determines the concentration of the disinfecting agent and the amount of exposure time. Generally, the less concentrated solution for the shortest time interval to obtain clean explants is desirable, as this will do the least amount of damage to the explant. The disinfecting agent can damage the explant, and the damaged tissue has to be removed before culturing the explant.

Disinfectant Agents

- *Sodium hypochlorite* (NaOCl), found in laundry bleach, is approximately a 5.25% (v/v) solution. Generally the laundry bleach is diluted in water to 5–25% (v/v), with 2 drops of Tween-20 added/100 ml. The duration of treatment is usually from 5 to 30 min followed by three to five rinses with sterile water. Commercial bleach solutions need to be made up fresh prior to use as the chlorine gas dissipates. It is best to use small containers of bleach and record on the bottle when it was opened. Over time the bleach will evolve chlorine gas and lose its effectiveness. A tip from David Constantine (Constantine Consulting & Agrotrading, drc@globalnet.co.uk via Plant-TC@ tc.umn.edu) is to adjust the pH of the bleach to about pH 7. The procedure is to freshly prepare the diluted bleach and adjust to pH 7 using 5 M HCl. The chlorine will come off more rapidly at the lower pH.
- *Calcium hypochlorite* (Ca(OCl)$_2$) at 0.8% (w/v) (8 g calcium hypochlorite in 1 liter water, stir for 10–5 min, allow to settle, decant solution, and mix with equal parts of water), add 2 drops of Tween-20/100 ml, treat tissue for 5–30 min.
- *Ethyl or isopropyl alcohol* at 70% (v/v) can be used to swab the plant material prior to disinfestation, and it can be used to dip the plant material in for 1–5 min before or after sodium or calcium hypochlorite treatment.

- *Hydrogen peroxide* (H_2O_2), a powerful oxidant, can be used at 3–10% (v/v) for 1–30 min prior to sterile water rinse in combination with other disinfestants or alone. An interaction between NaOCl and H_2O_2 is toxic to the plant tissue; therefore it is important to rinse the explant thoroughly if both are used.
- *Chlorine gas* (Cl_2) is effective for dry seed disinfestation. Place dry seeds in a petri dish inside a vacuum desiccator (this should be done in a fume hood). Partially open the top of the Petri dish and put 100 ml of Clorox bleach in a beaker inside the desiccator. Open the top to the desiccator a crack and slowly pipette 3.3 ml of concentrated HCl into the Clorox, being careful so that the HCl forms a layer on top of the Clorox (this allows slow mixing) and immediately seal the top of the desiccator. Chlorine gas will be generated. About 16–18 h later, carefully open the dessicator and use a glass rod to scoot the top onto the Petri dish. Close the front of the fume hood again and let the chlorine gas dissipate for a couple of hours. Seal the Petri dish with parafilm and store the sterile seeds until use.
- *Sodium dichloroisocyanurate* (*dichloroisocyanuric acid sodium salt*) (NaDCC) may be less toxic to plant tissues that are sensitive to the sodium and calcium hypochlorite and does not need to be rinsed off.
- *Isothiazolone biocide* (PPM) contains methyliosthizole and chloromethylisothiazole (Niedz, 1998) and is a product sold by Plant Cell Technology, Inc. (Washington, D.C.) which is advertised to ". . . effectively prevent and reduce microbial contamination in your plant tissue cultures." It is used as a pretissue soak as well as incorporated into the culture medium. It is reported to be effective, but at high concentrations it can be toxic to plant tissue. Miyazaki *et al.* (2011) used PPM under vacuum infiltration to eradicate bacterial endophytes from callus and subsequent shoots from the callus. It has been shown to be effective in also controlling endophytic bacteria (Miyazaki *et al.*, 2010).
- *Antibiotics* (*gentamicin and ampicillin*) may be beneficial to cut down on explant contamination after disinfestation using ethanol and bleach. A 50- to 100-mg/liter antibiotic solution used as a 30-min tissue soak prior to culture may eliminate some microorganisms. Seeds and embryos and *in vitro* cultures of ash had numerous bacterial contaminants (Donnarumma *et al.*, 2011). The authors used ampicillin to limit bacterial growth during culture, and the antibiotic had negligible phytotoxic effects on the explant, whereas tetracycline had phytotoxic effects.
- *Mercuric chloride* ($HgCl_2$) is a very hazadorous chemical that is often reported to be used as a topical explant sterilant. Thompson *et al.* (2009) compared mercuric chloride at 0.2% and household bleach at a 1:3 dilution in water. They found both were similar in elimination of contaminants, but household bleach resulted in better shoot development. These authors also pointed out the hazard of using mercuric chloride.

There is simply no justification for using this compound as a surface sterilant; however, the current literature has way too many references citing this compound's use. According to the Material Safety Data Sheet (MSDS), found on the internet, mercuric chloride should only be handled by personnel protected by splash goggles, synthetic aprons, vapor and dust respirators, and gloves. Mecuric chloride has chronic effects on humans, including possible action as a carcinogenic compound, being mutagenic to mammalian somatic cells, a toxin in the female reproductive system, and causing damage to the brain, peripheral and central nervous system, the skin, and eyes. It is very hazardous in the case of skin contact and is corrosive to eyes, and lungs if inhaled; immediate medical attention is required. Safe disposal of the compound once it has been used is yet another major concern.

Explant Source

The season of the year, where plant material is being grown (growth chamber, greenhouse, field), and location of the explant on the source plant are often significant factors in the establishment of clean cultures. Plant material that is in an active state of growth such as spring flush shoot tip growth is generally always cleaner as compared to dormant shoot tissue. Plant material from the field is often more contaminated as compared to greenhouse or growth chamber-grown plant material. Sometimes taking potted plant material out of the greenhouse (high humidity) and placing it in the laboratory or office (a dryer environment) a few weeks prior to taking explant material from the plant can help reduce contamination. Plant material growing in the soil (roots, tubers, bulbs) or near the soil surface (stolons, rhizomes, orchid protocorms, shoots from rosettes) is usually harder to clean than aerial plant material. Also refer to Bunn and Tan (2002).

A suggestion from Dr. Mike Kane (Environmental Horticulture Dept., Univ. of Florida via Plant-tc@tc.umn.edu) to obtain clean explants from rhizomes of aquatic plants which are extremely challenging to clean, is to divide the rhizome into multinodal segments and to treat in 50–100 mg/liter gibberellic acid to promote bud growth. The 1- to 2-mm buds, which develop from the rhizome, can be used as the explant source.

Internal Microbial Contamination

It is not uncommon that the explant material can harbor internal microorganisms (Armstrong, 1973; Tabor et al., 1991; Misaghi & Donndelinger, 1990; Petit et al., 1968; Gagne et al., 1987; Philipson & Blair, 1957; Lukezic, 1979; Sanford, 1948; DeBoer & Copeman, 1974; Hollis, 1951; Sturdy & Cole, 1974; Meneley & Stanghellini, 1974, 1975; Samish et al., 1963; Bugbee et al., 1987; Tervet & Hollis, 1948; Knauss & Miller, 1978; DePrest & Van Vaerenbergh, 1980; Schreiber et al., 1996). When this is the case, it can be very difficult to establish clean cultures. Explants from rapidly growing shoot tips, ovules of immature fruit, flower parts both mature and immature, and runner tips are usually least likely to harbor internal contaminants. Bulbs, slow-growing shoots or dormant buds, roots, corms, and underground rhizomes can have a heavy load of external and internal contaminants. Seeds can be aseptically germinated to provide clean explants from the root, hypocotyl (seedling stem), cotyledon, and shoot.

Interestingly, Abreu-Tarazi *et al.* (2010) reported that five-year-old *in vitro* propagated pineapple plants showing no signs of contamination had endophytic microorganisms (*Actinobacteria, Alphaproteobacteria* and *Petaproteobacteria*) in the roots and leaf tissue. These organisms were detected using the molecular techniques of denaturing gradient gel electrophoresis (DGGE) and polymerase chain reaction (PCR). Likewise, Ulrich *et al.* (2008) reported that poplar, larch and spruce cell cultures in culture for at least five years had a high density of endophytic bacteria mostly in the genus *Paenibacillus*, and some had *Methylobacterium, Stenotrophomonas*, and *Bacillus*.

The use of antibiotics or fungicides in the nutrient medium is generally not successful. Antibiotics also need to be filter sterilized and added to cooled media as they are heat labile. These agents can repress the growth of some microorganisms, but when the antibiotic or fungicide is removed the microorganism will generally reappear. These agents can also suppress the plant tissue or even kill it. The use of antibiotics as selectable markers in plant transformation is based on the fact that the antibiotic is lethal to the nontransformed tissue and only transgenic tissue containing the gene for antibiotic resistance will survive and grow. Sometimes if the identity of the microorganism is known such as when one is doing transformation experiments using *Agrobacterium tumefaciens*, one can effectively use clavamox, augmentin, carbenicillin, and other antibiotics to eliminate or control the *Agrobacterium* from overgrowing the explant.

Thurston *et al.* (1979) did an extensive study on the phytotoxicity of fungicides and bactericides in orchid culture, and they did find some combinations that were effective in controlling contamination. Pollock *et al.* (1983) also reported on screening over 20 antibiotics to evaluate their toxicity on tobacco protoplasts.

Explant Pretreatment to Control Microbial Contaminants

Heat treatments can be effective for some viruses as well as fungi and bacteria. Bulbs can sometimes be freed of systemic fungal (*Fusarium* sp.) contamination by giving the bulb a heat treatment in a hot water bath for 1 h at 54°C (Hol & van der Linde, 1992). Potted plants can be put in a growth chamber with the temperature at 35–40°C for prescribed time periods in order to rid some plants like potatoes of viruses. A fungicide treatment (benlate or benomyl at 0.5%) of plant material prior to culture may help reduce fungal contamination.

PERSONNEL

In order to minimize introducing contamination from individuals doing aseptic transfers, it is important to wash hands, fingernails, and arms with warm soapy water with a fingernail brush. If an individual is not going to wear sterile, latex gloves, one should spray the hands and arms with 70% alcohol, being extremely careful not to go near a flame until the hands and arms have air dried. Hair nets, masks, and a clean laboratory coat are beneficial. Personnel should not work in the greenhouse or field prior to working in the laminar air flow hood, as mites and other contaminants can be carried in on clothing and hair. Work in the aseptic transfer hood, not at the outer edge, and minimize talking when working in the transfer hood. Also try not to lean over the open culture container or Petri dish where the explant is being dissected.

LAMINAR AIR FLOW HOOD

A laminar air flow hood is extremely useful for aseptic transfer work. A laminar flow hood creates a positive air pressure toward the worker and creates a flow of clean air over the work area. High efficiency particulate air (HEPA) filters are "absolute" or 99.97% free of 0.3-μm particles. There are prefilters to filter out large particles either at the base of the hood or on top of the hood. Prefilters should be cleaned on a regular schedule depending on the amount of dust in the laboratory to maintain the life of the expensive HEPA filters. Wet-mop the laboratory and avoid carpeting or use of vacuum cleaners.

INSECTS

One of the biggest problems in a tissue culture laboratory is to have an insect infestation, especially mites. There are several different mites, *Dermataphagoides pteronyssimus*, *Dermatafagoides farinae*, and *Tyrophagus putrescentiae* or dust mites, which can be a major problem. Mites can live in air-conditioning duct work and filters and are found on clothes and hair. Mites can move in air currents. Generally, mites are more active in early evening and are attracted to moisture and organic material. They feed on fungi, not plant material or callus. The mite's life cycle is about 2 weeks. The cultures are contaminated due to the mite traveling in and out of the culture vessels without being noticed until a few days later one sees their tracks, which are visible by the path of fungal contamination which is spread by the mites[1] legs.

Insects can be introduced into the lab by personnel (on hair, clothing, shoes, etc.), potted plant material, large explant material, adjacent laboratories working on insects, or storing large quantities of seed or dry plant material. The air-conditioning duct systems can be contaminated, especially with fungal organisms.

Control of Insect Contamination

Wrapping or sealing cultures is a very tedious but effective means of preventing insects from entering and sealing off insects that are already present in a culture vessel. Sometimes cultures in Petri dishes are placed in plastic boxes or Ziplock baggies. Sealing culture vessels must be done if an insect problem is in progress to control it. Contaminated cultures must be autoclaved immediately. Parafilm is commonly used to wrap culture vessels. A discussion in the Plant-tc listserver mentions the use of Glad-Wrap, which is a polyethylene film with good gas permeability for oxygen, carbon dioxide, and ethylene and with a low permeability for water vapor as compared to Saran-Wrap, a polyvynilidene copolymer, which has low gas permeability. Any type of wrap can create problems in data replication due to variances in the number of layers of wrap and tightness of the wrap. These parameters will make the gas exchange ratios variable. Containers that can be closed by membrane-vented lids allow repeatable gas exchange and exclude insects. Membrane-vented lid closures are available from Sigma, Gibco-BRL, and Osmotek.

Insecticides are effective in killing the insects and numerous strategies have recently been discussed in the Plant-tc listserve. Some of the suggestions from this and other sources are as follows:

- Aerosol fly spray with pyrethrums like Raid can be sprayed on Friday afternoons for about 6 weeks. Pay particular attention to the corners of the room which usually accumulate dust where mites "hang out." The lab should be vented on Monday before personnel reenter the laboratory. Thereafter, the lab should be sprayed once a month to control reintroduction of insects (Dave Kirk, Tauranga, New Zealand).
- Benzyl-benzoate spray is also effective for 3 months.
- Using acaricides like fenbutatin oxide and difocol. Difocol can be added to the culture medium or used to saturate cotton plugs. The saturated cotton plugs can be placed in the culture medium with the explants. Additionally, paper soaked in these chemicals can be placed on the culture shelves (J. Pype, Lab for Horticulture, University of Gent, Belgium).
- Spray with an acaricide once a week.
- Pest strips can be hung in the culture room; also, use shelving paper that is impregnated with an insecticide.
- Base board can be painted with an insecticide.
- Sticky carpets at doors remove dust and dirt on shoes and trap migrating insects.

MEDIA

The tissue culture media is generally autoclaved at 121°C (21 psi) for 15 min. This is adequate for culture vessels filled with 10–1000 ml of liquid medium. A discussion regarding the effects of autoclaving on various components in the nutrient media is in the chapter on media preparation.

Some media components are heat labile. These materials have to be filter sterilized and added to the nutrient media after it has been autoclaved and cool enough to handle. Membrane filters with a pore size of 0.22 μ are recommended. It is important to thoroughly mix the media after addition of the filter-sterilized component before distributing the media to the sterile culture vessels. A description of how this can be done is included in Chapter 3.

INSTRUMENTS

Common instruments used in preparing explants for culture are forceps of varying lengths and scalpels with varying blade sizes. Instrument packages can be wrapped in aluminum foil or placed in a stainless steel instrument tray and autoclaved at 121°C for 15–20 min or placed in a dry oven at 140–160°C for 4 h. During use of the instruments in the culture hood there are several methods used to keep the instruments sterile:

- Place the instrument in a test tube that has 95% alcohol with about one-third of the instrument not submerged in the alcohol. The test tube should not be in a rack because if the bottom of the test tube should break, alcohol will rapidly spread, ignite, and cause a fire. Cheesecloth placed at the bottom of the vessel can also help to prevent breakage of the alcohol container. The test tube should be placed in a sturdy flask or beaker. Water at the bottom of the container holding the test tube helps to prevent the container from tipping over. The instrument is then flamed and allowed to cool before using. In flaming the instrument, if the individual elevates the instrument above the fingers holding the instrument, the burning alcohol will run down onto fingers, causing a burn. *Alcohol disinfestation of instruments can be dangerous, and one must be extremely careful to separate the alcohol-containing test tube on one side of the transfer hood from the flame on the other side of the hood. NEVER place the alcohol or Bunsen burner next to the container holding the alcohol. Never leave a beginning student alone in the transfer hood; the student must be properly supervised at all times. Always be sure the instrument has finished burning and cooled before putting it back in the alcohol.*
- Streiber *et al.* (1996) reported that *Bacillus macerans*, a bacterial contaminant, could be viable on forceps after being stored in 95% ethanol for several weeks. The bacteria even remained viable after flaming. The bacteria is eliminated only by autoclaving at 121°C for 20 min or by heating for 6–8 s over a Bunsen burner. Additionally, *Clavibacter* may survive alcohol flaming. For these reasons, washing the instruments to remove plant material on the surface and periodic autoclaving are desirable.
- One suggestion (Dan Cohen, Plant-tc listserver, MT Albert, New Zealand) is to adjust the pH of a sodium hypochlorite in the following manner: prepare a 10% solution of K_2HPO_4 (dipotassium hydrogen phosphate), add 2 ml of bleach (approximately 5% sodium hypochlorite, 5% available chlorine) to 96 ml of distilled water, and add 2 ml of 10% K_2HO_4, which results in 100 ml of hypochlorite buffered to pH 7 in a sodium phosphate buffer. This is toxic to cut plant surfaces and delicate plant material; however, it is effective in killing *Bacillus* spores at 1000 ppm for 30 s on instruments. They put instruments in this solution, dip in ethanol, and flame to dry.

- Bead sterilizers (Inotech Biosystems International, P.O. Box 21064, Lansing, MI 48909, 800/635–4070) are an excellent, safe option for instrument sterilization. The instrument is placed in the sterilizer ~10 s at 240°C. Beads are cleaned two to three times a year depending on use. Some workers indicate that blades dull faster in a bead sterilizer compared to in alcohol.
- Bacticinerators are also efficient in instrument sterilization. Instrument life is reduced due to the heat, causing warping. One must be very careful in removing the instrument as the handles can become very hot and cause severe burns.

ROOM AIR HANDLING SYSTEM

Duct work and filters on air-conditioning systems can become a source of insect and microbial contamination. Air handling ducts should be cleaned and properly maintained and prefilters venting the air into the culture room should be replaced frequently.

REFERENCES

Abreu-Tarazi, M. F., Navarrete, A. A., Andreote, F. D., Almeida, C. V., Tsai, S. M., & Almeida, M. (2010). Endophytic bacteria in long-term in vitro cultivated "axenic" pineapple microplants revealed by PCR-DGGE. *World Journal of Microbiology & Biotechnology*, *26*(3), 555–560.
Armstrong, D. (1973). Contamination of tissue culture by bacteria and fungi. In J. Fogh (Ed.), *Contamination in tissue culture* (pp. 51–64). New York: Academic Press.
Bugbee, W. M., Gudmestad, N. C., Secor, G. A., & Nolte, P. (1987). Sugar beet as a symptomless host for Corynebacterium sepedonicum. *Phytopathology*, *77*, 765–770.
Bunn, E., & Tan, B. H. (2002). Microbial contaminants in plant tissue culture propagating. In K. Sivasithamparam, & K. W. Dixon (Eds.), *Microorganisms in plant conservation and biodiversity* (pp. 307–335). Dordrecht, The Netherlands: Kluwer Academic Publishers.
DeBoer, S. H., & Copeman, R. J. (1974). Endophytic bacterial flora in *Solanum tuberosum* and its significance in bacterial ring rot diagnosis. *Canadian Journal of Plant Science*, *54*, 115–122.
DePrest, G., Van Vaerenbergh, J., & Welvaert, W. (1980). Infections bacteriennes dans des cultures in vitro. *Med. Fac. Landbouw Rijksuniv Gent.*, *45*(2), 399–410.
Donnarumma, F., Capuana, M., Vettori, C., Petrini, G., Giannini, R., Indorato, C., & Mastromei, G. (2011). Isolation and characterization of bacterial colonies from seeds and in vitro cultures of *Fraxinus* spp. From Italian sites. *Plant Biology*, *13*(1), 169–176.
Gagne, S. R., Rousseau, H., & Anton, H. (1987). Xylem-residing bacteria in alfalfa roots. *Canadian Journal of Microbiology*, *33*, 99–100.
Hol, G. M., & Van der Linde, P. C. G. (1992). Reduction of contamination in bulb-explant cultures of *Narcissus* by a hot-water treatment of parent bulbs. *Plant Cell Tissue & Organ Culture*, *31*, 75–79.
Hollis, J. P. (1951). Bacteria in healthy potato tissue. *Phytopathology*, *41*, 350–366.
Knauss, J. F., & Miller, J. W. (1978). A contaminant, *Erwinia cartovora*, affecting commercial plant tissue cultures. *In Vitro*, *14*(9), 754–756.
Lukezic, F. L. (1979). *Pseudomonas corrugata*, a pathogen of tomato, isolated from symptomless alfalfa root. *Phytopathology*, *69*, 27–31.
Meneley, J. C., & Stanghellini, M. E. (1974). Detection of enteric bacteria within locular tissue of healthy cucumbers. *Journal of Food Science*, *39*, 1267–1268.

Meneley, J. C., & Stanghellini, M. E. (1975). Establishment of inactive population of *Erwinia cartovora* in healthy cucumber fruit. *Phytopathology, 65*, 670–673.

Misaghi, I. J., & Donndelinger, C. R. (1990). Endophytic bacteria in symptom-free-cotton plants. *Phytopathology, 80*, 808–811.

Miyazaki, J., Tan, B. H., & Errington, S. G. (2010). Eradication of endophytic bacteria via treatment for axillary buds of Petunia hybrida using plant preservative mixture (PPM™). *Plant Cell, Tissue & Organ Culture, 102*, 365–372.

Miyazaki, J., Tan, B. H., Errington, S. G., & Kuo, J. J. S. (2011). *Macropidia fuliginosa*: its localization and eradication from in vitro cultured basal stem cells. *Australian J. Bot., 59*(4), 363–368.

Niedz, R. P. (1998). Using isothiazole biocides to control microbial and fungal contaminants in plant tissue cultures. *HortTechnology, 8*, 598–601.

Petit, R. E., Taber, R. A., & Foster, B. G. (1968). Occurrence of *Bacillus subtilis* in peanut kernels. *Phytopathology, 58*, 254–255.

Philipson, M. N., & Blair, I. D. (1957). Bacteria in clover root tissue. *Canadian Journal of Microbiology, 3*, 125–129.

Pollock, K., Barfield, D. G., & Shields, R. (1983). The toxicity of antibiotics to plant cell cultures. *Plant Cell Reports, 2*, 36–39.

Samish, Z., Etinger-Tutczynska, R., & Bick, M. (1963). The microflora within the tissue of fruits and vegetables. *Journal of Food Science, 28*, 259–266.

Sanford, G. B. (1948). The occurrence of bacteria in normal potato plants and legumes. *Science and Agriculture, 28*, 21–24.

Schreiber, L. R., Domir, S. C., & Gingas, V. M. (1996). Identification and control of bacterial contamination in callus cultures of *Ulmus americana*. *Journal of Environmental Horticulture, 14*(2), 50–52.

Sturdy, M. L., & Cole, A. L. J. (1974). Studies on endogenous bacteria in potato tubers infected by *Phytophthora erythroseptica* hybr. *Annals of Botany, 8*, 121–127.

Taber, R. A., Thielen, M. A., Falkinhan, J. O., III, & Smith, R. H. (1991). *Mycobacterium scrofulaceum*: A bacterial contaminant in plant tissue culture. *Plant Science, 78*, 231–236.

Tervet, J. W., & Hollis, J. P. (1948). Bacteria in the storage organs of healthy plants. *Phytopathology, 38*, 960–962.

Thompson, I. M., Laing, M. D., & Beck-Pay, S. L. (2009). Screening of topical sterilants for shoot apex culture of *Acacia mearnsii*. *Southern Forests, 71*(1), 37–40.

Thurston, K. C., Spencer, S. J., & Arditti, J. (1979). Phytotoxicity of fungicides and bactericides in orchid culture media. *American Journal of Botany, 66*(7), 825–835.

Ulrich, K., Stauber, T., & Ewald, D. (2008). *Paenibacillus*—a predominant endophytic bacterium colonizing tissue cultures of woody plants. *Plant Cell Tissue & Organ Culture, 93*(3), 347–351.

Chapter 6

Callus Induction

Chapter Outline

CALLUS INITIATION

As a first step in many tissue culture experiments, it is necessary to induce callus formation from the primary explant. This explant may be an aseptically germinated seedling or surface-sterilized roots, stems, leaves, or reproductive structures. Callus is a wound tissue produced in response to injury. The callus is a proliferation of cells from the wounded or cut region of an explant. Callus is generally made up of friable, large, vacuolated cells that are highly differentiated, but are unorganized. Callus can be hard and compact, and can contain regions of small meristematic cell clusters. It is generally the meristematic, undifferentiated cells that are competent to

Plant Tissue Culture. Third Edition. DOI: 10.1016/B978-0-12-415920-4.00006-2

regenerate via somatic embryo or organ initiation (usually shoot or root development). Not all cells in an explant contribute to the formation of callus and, more
importantly, certain callus cell types are competent to regenerate organized structures. Other callus cell types do not appear to be competent to express totipotency.

A recent article by Wang *et al.* (2011) histologically examined callus induction from alfalfa leaf explants. They showed callus initiation from the cut leaf
surface and veins. Interestingly, callus cells from procambial (vein) cells rarely
developed somatic embryos; in many plant species totipotent cells do originate
from procambial cells. Alfalfa leaf mesophyll cells dedifferentiated to form
somatic embryos. Early microscopic visual selection is usually necessary to
select for the cell type that is regenerable. A recent publication by Naor *et al.*
(2011) provides additional insight into obtaining callus from field-grown grape-
vine-inverted (placed in the medium with the top of the explant in the medium
and the basal end up) nodal stem section explants without added plant growth
regulators. The explant itself initiated callus proliferation presumably as a result
of basipetal auxin movement in the stem explant.

The level of plant growth regulators (auxin, cytokinins, giberellins, ethylene,
etc.) is a major factor that controls callus formation in the culture medium. Concentrations of the plant growth regulators can vary for each plant species and can
even depend on the source of the explant or individual plant genotype, age, nutritional status, etc. Culture conditions (temperature, solid media vs agar solidified,
light, etc.) are also important in callus formation and development. An examination of the literature between 2007 and 2011 confirms that there is no universal
method to successfully obtain callus cultures from all plant species. There are
thousands of journal articles that describe research using varied explants, culture
media, plant growth regulator levels and combinations, as well as other addendum
to the culture medium, and varied culture conditions to induce callus and regenerable callus from specific plant species. Some recent publications illustrating current protocols to evaluate these parameters include: Garcia *et al.* (2011); Dhar &
Joshi (2005); Gao *et al.* (2010); Irvani *et al.* (2010). The exercises in this chapter
provide experience in varied techniques for using different explants, species and
culture conditions to observe and study callus induction and gain experience.

Once established, callus cultures may be used for a variety of experiments.
The callus cultures in these chapters will be used to study protoplast isolation,
cell type, cellular selection, somatic embryogenesis, organogenesis, and secondary product production. Additionally, regenerable callus is useful as a target
for genetic transformation.

Purpose: To gain experience in aseptic technique and callus induction from
varied explants (seedlings, fruit, inflorescence, root).

Medium Preparation: 1 liter equivalent, Callus Initiation Medium.

1. Into a 2000-ml Erlenmeyer flask pour 500 ml of deionized, distilled water.
2. Mix in the following:
   a. 10 ml each Murashige and Skoog salts: nitrates, halides, sulfates,
      NaFeEDTA, PBMo

    b. 10 ml thiamine stock (40 mg/liter)
    c. 10 ml myo-inositol stock (10 g/liter)
    d. 30 g sucrose
    e. 1.0 ml kinetin stock (10 mg/100 ml)
    f. 3.0 ml 2,4-D stock (10 mg/100 ml)
    g. 10 ml vitamin stock

3. Adjust volume to 1000 ml. Adjust pH to 5.7.
4. Add 8 g/liter TC agar or Difco-Bacto agar. Cap Erlenmeyer flask with aluminum foil.
5. Autoclave for 15 min at 121°C, 15 psi.
6. Distribute 25 ml per sterile plastic petri dish (100 × 20 mm) using a transfer hood.

Explant Preparation

Carrot Seedlings

1. Remove aseptically germinated seedlings and place in a sterile Petri dish (see Chapter 4).
2. Cut up seedling and culture leaf, stem, and root tissue. The seeds that did not produce small plants can still be used if the radicle (root) is protruding. Remove the seed coat and slice the seed into several pieces.

Lemon

1. Green immature fruit will yield better results than ripe fruit. Wash the fruit in warm, soapy water. Sever the stem and calyx tissue that may be attached to the fruit; remove necrotic tissue in the rind.
2. Cut the fruit into ½-inch sections.
3. Disinfect in 15% chlorine bleach (+ Tween-20) for 15 min. Rinse three times in sterile water. The midsections of the fruit may provide better explants. Culture clumps and individual juice vesicles.

Broccoli

1. Wash florets in warm, soapy water.
2. Disinfect in 15% chlorine bleach (+Tween-20) for 15 min. Rinse three times in sterile water. Cut off tissue burned by the chlorine bleach. Plant sections of the floret, individual flowers, and peduncle tissue.

Carrot Root Tissue

Use fresh carrots with green tops still attached.

1. Scrub root in warm soapy water using a brush. Cut out bad spots on the root.
2. Cut the root into ½-inch sections and surface sterilize in 15% chlorine bleach for 15 min. Rinse three times in sterile water. Cut off tissue burned by chlorine bleach and culture.

Culture

Place all cultures in the dark (a drawer or cabinet will work) at about 27–30°C; record the temperature.

Observations

Students are responsible for observations once a week over a 6-week period. Observations should include notes on culture conditions, callus formation, and contamination. It is very important to observe cultures under a dissecting microscope once a week and to record observations and make drawings. Callus will arise from different regions of the explant, such as the cambial tissue of the carrot root. Additionally, large numbers of somatic embryos can form directly from the wounded carrot seed explants. Many of these activities are not visible to the naked eye, particularly early cell divisions and explant enlargement. At the end of 6 weeks, students can sacrifice the cultures and observe cell and callus types by making wet mounts on slides and observing under a light microscope.

A discussion should be included as part of the notebook entry. This discussion should include comments on callus formation in response to explant source, why callus formed or why it did not, and comparisons of the responses of different tissue sources such as seed derived callus vs mature tissue. Illustrations of callus and cell type should be reported in the observation section.

The carrot callus cultures can be used to initiate suspension cultures for somatic embryogenesis, growth curves, and salt selection studies. Additionally, carrot callus can be used to isolate pure lines of different pigmented cell lines.

Callus from broccoli and lemon can be the starting point for special student projects.

Data and Questions

1. Observe and record the weekly progress of the explants. Determine where on the explant callus formation is apparent. Is the entire explant involved in callus formation? Do all explants respond in a similar fashion? Are the callus cells all identical in appearance?
2. Express the plant growth regulator concentrations in the medium in parts per million.
3. What is the molar concentration of the kinetin and 2,4-D stock solutions, and what is the medium concentration of these plant growth regulators?
4. At the time specified by the instructor (4–6 weeks) remove some callus and prepare a wet-mount on a slide and observe the cells under a microscope. Draw the cells. Note differences in the cell sizes and shapes.

EXPLANT ORIENTATION

Endogenous levels of plant growth regulators and polar plant hormone transport within the explant can influence callus induction from your explant. Many times when attempts are made to duplicate prior published work, small details such as explant preparation and position on the medium have been left out of the materials and methods section, resulting in different responses. For some plants this is not critical; however, for other plants it is. For example, a methods section may indicate "Hypocotyl sections were placed in culture." This exercise will illustrate the varying response of hypocotyl and/or cotyledon explants depending on how they are sectioned and placed in the medium. Similar variance in explant response can be expected from other explants.

Purpose: To demonstrate the effect of explant orientation on callus induction from cotton and sunflower seedling explants.

Medium Preparation: 1 liter equivalent, Cotton Explant Medium.

1. Into a 2000-ml Erlenmeyer flask pour 500 ml of deionized, distilled water.
2. Mix in the following:
   a. 10 ml each Murashige and Skoog salts: nitrates, halides, NaFeEDTA, sulfates, and PBMo
   b. 10 ml thiamine stock (40 mg/liter)
   c. 10 ml myo-inositol stock (10 g/liter)
   d. 30 g glucose
   e. 10 ml NAA stock (10 mg/100 ml)
   f. 10 ml 2iP (10 mg/100 ml)
3. Adjust volume to 1000 ml. Adjust pH to 5.7.
4. Add 8 g TC agar or Difco-Bacto agar. Cap with aluminum foil.
5. Autoclave for 15 min at 121°C, 15 psi.
6. Distribute 25 ml per sterile plastic Petri dish (100 × 20 ml).

For the sunflower explant, use the Callus Initiation Medium described at the beginning of this chapter.

Explant

Cut hypocotyl and/or cotyledon sections from the aseptically germinated cotton or sunflower seeds that were germinated in the "Aseptic Germination of Seeds" exercise. The method for cutting hypocotyl sections is shown in Fig. 6.1. Culture the explants on the culture shelf.

Experiments can be designed to ask the questions:

- If the explants are cut into disks and placed in the medium right side up vs upside down, are there differences in time of callus induction and amount of callus induction from hypocotyl sections near the leaf vs the root?

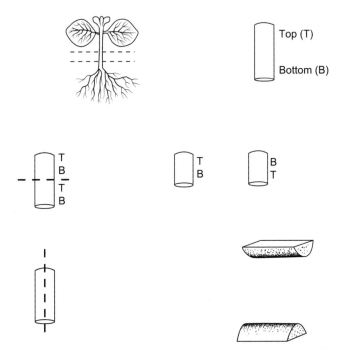

FIGURE 6.1 To obtain seedling hypocotyl sections, cut 1-cm sections (top of figure). Cross sections are placed with top up and bottom down and with bottom up and top down, respectively (center of figure). Longitudinal sections are placed with cut surface up and with cut surface down (bottom of figure).

- Are there any differences in amount and time of callus induction if the explants are placed upside down in the medium?
- Is there a difference in callus induction from cotyledon sections if they are placed right side up or upside down on the culture medium?
- Do cotyledons produce more callus than hypocotyl explants?
- Is the callus identical in appearance?

Use your imagination to set up experiments to test other ideas (total dark vs light: dark culture; temperature; half MS vs full MS; etc.) in regard to explant orientation and callus response.

Observations

Observations (each week for 4 weeks) should include drawings of the explants indicating the areas and relative amounts of callus production. Observe the cultures under a dissection microscope. These cultures can be a starting point for the isolation of pure pigmented cell lines after 6–8 weeks in culture.

ESTABLISHMENT OF COMPETENT CEREAL CELL CULTURES

Historically, the establishment of competent cell cultures from monocotyledons has been difficult. In the late 1970s and 1980s many workers reported a strong effect of genotype on culture regeneration ability. Some workers have established that regeneration from maize cultures was genetically controlled by nuclear genes (Hodges *et al.*, 1986; Tomes & Smith, 1985). Peng and Hodges (1989) presented evidence that rice regeneration was under the control of both nuclear and cytoplasmic genes. The ability to form regenerable callus in sorghum varied among genotypes, was heritable, and acted as a dominant trait (Ma *et al.*, 1987), which implied certain cultivars lacked the genes for regeneration in culture. The problem with the concept that certain cultivars lack the genes for regeneration is that a cultivar that was established as nonregenerable in one laboratory was later reported to be regenerable in other laboratories, ruling out the idea that some cultivars lack the genetic information to form regenerable cultures (Bhaskaran & Smith, 1990).

It seems more reasonable that nuclear genes are involved in the control of cultivar responsiveness to plant growth regulator type and concentration in the culture medium (Close & Gallagher-Ludeman, 1989). Cultivar differences, then, are probably related to variations in endogenous hormone levels, which are genetically established. Explants (i.e., leaves, roots, anthers, stems, etc.) from a single cultivar, even a plant seedling, do not respond identically in culture on the same medium. Again, this is most likely because of varying gradients in endogenous hormones (Wernicke & Brettell, 1982) and cell types (meristematic vs highly differentiated) within the explant.

Cereal callus cultures must be initiated from explants that contain meristematic cells. These explants include immature embryos collected from flowering plants and mature seeds or seedling parts, including the shoot meristem and basal sections of young leaves. A second important factor is early visual selection of the compact, nodulated callus type that is milky white to yellow in color (Bhaskaran & Smith, 1988; Heyser & Nabors, 1982). The nonembryogenic callus is usually loose, crystalline, yellow to brown in color (Abe & Futsuhara, 1985; Nabors *et al.*, 1983), and faster growing and will overgrow and cover the embryogenic callus.

Purpose: To visually identify and select the embryogenic callus type from rice on callus induction medium (Peterson & Smith, 1991).

Medium Preparation: 1 liter equivalent, Callus Induction Medium.

1. Into a 2000-ml Erlenmeyer flask pour 500 ml of deionized, distilled water.
2. Mix in the following:
   a. 10 ml each Murashige and Skoog salts: nitrates, halides, NaFeEDTA, sulfates, and PBMo
   b. 10 ml thiamine stock (40 mg/liter)
   c. 10 ml myo-inositol stock (10 g/liter)
   d. 30 g sucrose

  e. 10 ml vitamin stock (see Chapter 3)
  f. 35 ml 2,4-D stock (10 mg/100 ml)
3. Adjust volume to 1000 ml. Adjust pH to 5.7.
4. Add 4 g Sigma type 1 agarose. Cap with aluminum foil.
5. Autoclave for 15 min at 121°C, 15 psi.
6. Distribute 25 ml per sterile plastic Petri dish (100 × 20 ml).

Explants

Use mature seeds of *Oryza sativa* L. Texas cultivar "Lemont." (An experimental variation is to examine other cultivars and compare responses.) Dehusk the seeds and surface sterilize in 70% ethanol for 1 min followed by 30% chlorine bleach for 30 min and five rinses in sterile water.

Plant the seeds on the Callus Induction Medium. Culture in the dark at 27–30°C. At 2–3 weeks separate the embryonic callus, which is smooth, white, and knobby, from the yellow-to-translucent, wet, crystalline-appearing, non-embryogenic callus using a dissecting microscope in the culture hood. Use this material for the Regeneration of Rice exercise.

Data and Questions

1. Did all seeds germinate? What was the percentage? Was contamination a problem?
2. Sacrifice a culture, prepare a wet mount of the cells, and record your observations.
3. Did all cultures initiate the callus?
4. If you examined several rice cultivars, were there differences among them?

SALT SELECTION *IN VITRO*

Plant cells in culture can be a very useful experimental system for isolating and studying the cellular level response to various environmental stresses. Great care, however, must be exercised in analysis of this system because the whole-plant response to environmental stresses is a combination of many factors and cellular level activities may not be an accurate system from which to generalize back to the whole-plant system response.

In any event, cellular-level studies can be useful for studying the effects of ions (Na, Al, etc.), herbicides, pesticides, and pathogen-produced toxins. Rai *et al.* (2011) have reviewed the literature on *in vitro* cellular selection for salt tolerance using sodium chloride, disease resistance using fungal toxins, and abiotic stresses like drought, low temperature, metals, and ultraviolet radiation. The pros and cons and successes of these strategies are presented and extensive literature citations are included.

Several strategies can be employed in these types of studies. One is screening cells for tolerance to increasing levels of the compound and selecting cells with high levels of tolerance. These might be cells that will grow with high levels of the selective agent in the medium. In some studies, selection is done in a stepwise fashion, gradually increasing the selective agent in the medium at each subculture. If plants can then be regenerated, perhaps this cellular level trait will enhance their overall tolerance to that specific stress.

One can screen cells of different genotypes with known differences in heavy metal or salt tolerance to see if these differences are maintained in cell cultivar (Yeo & Flowers, 1983; Tal & Shannon, 1983). Many of these studies are focused on developing a better understanding of the control and regulation of important physiological responses at the cellular level.

Purpose: To screen carrot callus cultures for tolerance to NaCl.

Medium Preparation: 1 liter equivalent, Salt Selection Medium.

1. Into a 2000-ml Erlenmeyer flask pour 500 ml of deionized, distilled water.
2. Mix in the following:
   a. 10 ml each Murashige and Skoog salts: nitrates, halides, NaFeEDTA, sulfates, and PBMo
   b. 10 ml thiamine stock (40 mg/liter)
   c. 10 ml myo-inositol stock (10 g/liter)
   d. 30 g sucrose
   e. 2 mg glycine
   f. 10 ml vitamin stock (see Chapter 3)
   g. 3.0 ml 2,4-D stock (10 mg/100 ml)
3. Adjust volume to 800 ml. Divide into four parts of 200 ml each. Weigh out and add the following amounts of NaCl, respectively:
   a. 0 NaCl
   b. 0.2 g NaCl/liter = 0.05 g NaCl/250 ml
   c. 1.6 g NaCl/liter = 0.4 g NaCl/250 ml
   d. 6.0 g NaCl/liter = 1.5 g NaCl/250 ml
4. Adjust volume to 250 ml. Adjust pH to 5.7.
5. Add 1.5 g TC agar to each. Melt.
6. Distribute 25 ml per culture tube (25 × 150 mm). Cap using a color code for each treatment.
7. Autoclave for 15 min at 121°C, 15 psi.

Explants

Carrot callus was established in the "Callus Initiation" exercise. Each student should prepare 5 tubes of each treatment for a total of 20 tubes.

1. Weigh a 250 ± 2 mg callus piece to culture in each tube. Weigh additional pieces to use for an initial weight determination. If a balance in the transfer hood is not available, divide the pieces of callus as evenly as possible, take three for random initial wet and dry starting weights, and culture the rest.

2. Place the callus on the medium. Be careful that the same amount of each callus piece is in contact with the medium.
3. Label and weigh foil for wet and dry weight determination. Place the callus for dry weight measurement on the preweighed foil and place the foil/callus in the drying oven (48 h at 60°C).
4. At the end of 4 weeks, sacrifice all the cultures to measure wet and dry weights. Present data in table form and graph the wet and dry weights over the 4 weeks.

Data and Questions

1. Describe the appearance of the callus at each concentration of NaCl.
2. What NaCl level resulted in 50% survival?
3. Do you think the surviving cells would survive subculture with the same NaCl level or with a higher NaCl level?
4. Do wet and dry weights of callus correlate?
5. If the trait being selected was the result of several genes, would this be a viable scheme to obtain stress tolerant plants?

GROWTH CURVES

The rate of growth of callus tissue parallels in many ways the sigmoid curve seen in populations of single-celled organisms. There are usually five stages as shown in Fig. 6.2. The behavior of cells of callus tissue is different during each stage of growth. The medium can also influence how long the callus remains at a particular stage.

For many experimental procedures, it is necessary to use callus at a specific developmental point along the growth curve. When examining chromosomes, one finds the greatest number of cells in metaphase during the exponential

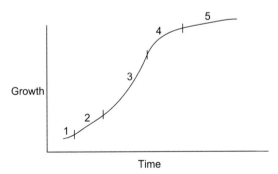

FIGURE 6.2 The rate of growth of callus tissue usually has five stages: (1) a lag phase in which cells prepare to divide; (2) a period of exponential growth in which cell division is maximal; (3) a period of linear growth in which division slows down and cells enlarge; (4) a period of decelerating growth; and (5) a stationary or no-growth period in which the number of cells is constant.

phase of growth in which there is rapid cell division and proliferation. Cell growth and development are most evident during the linear stage. Once the culture enters stage 4, decelerating growth, it should be transferred to fresh medium. In addition, when secondary product formation is examined, it is critical to determine the growth stage that yields the greatest quantity of the desired product.

Deceleration of growth is due to depletion of nutrients, drying of agar, production of toxic byproducts, and O_2 depletion in the interior of the callus. Generally only the healthy-appearing tissues are subcultured. However, observe the different cell types under a dissecting microscope in the transfer hood. Often the slow-growing, white to tan opaque cells are the important cell types to subculture to establish culture lines that will regenerate. Many times the green, rapidly growing cell lines do not regenerate.

A growth curve can be established by using several methods. The first is to sacrifice cultures at different time intervals and take wet and dry weights. Always replicate and average to establish a data point. Collection of wet weight data is rapid, and wet weights generally correlate with dry weights at fresh weight values above 0.5 g. A cell count can be made in which cells are treated with chromium trioxide (4%) at 70°C for 2–15 min. This mixture is agitated, and portions are counted on a hemocytometer slide. Growth curves for suspension cultures can be established by withdrawing 5- to 10-ml aliquots and centrifuging in a graduated conical test tube. The packed cell volume is measured. Suspension culture growth can also be established by collecting a 5- to 10-ml aliquot and collecting cells on preweighted filter paper in a Millipore filter holder with a vacuum. Wash the cells with water and place the filter paper in a preweighed Petri dish. Dry in an oven for 48 h at 60°C and weigh.

Cotton

Purpose: To subculture and establish a growth curve for cotton callus, *Gossypium hirsutum* L. (Price *et al.*, 1977). For experimental variation, compare different brands of agar or the effect of different brands of charcoal on cell growth. Many other factors such as vitamins, temperature, light intensity, and plant growth regulators can also be tested. Also see references for ideas.

Medium Preparation: 1 liter equivalent, Cotton Growth Curve Medium.

1. Into a 2000-ml Erlenmeyer flask pour 500 ml of deionized, distilled water.
2. Mix in the following:
   a. 10 ml each Murashige and Skoog salts: nitrates, halides, NaFeEDTA, sulfates, and PBMo
   b. 10 ml thiamine stock (40 mg/liter)
   c. 10 ml myo-inositol stock (10 g/liter)
   d. 30 g glucose
   e. 10 ml 2iP stock (10 mg/100 ml)
   f. 1.0 ml NAA stock (10 mg/100 ml)

3. Adjust volume to 1000 ml. Adjust pH to 5.7.
4. Add 8 g TC agar or Difco-Bacto agar. Melt.
5. Distribute 25 ml per culture tube (25 × 150 mm). Cap.
6. Autoclave for 15 min at 121°C, 15 psi. Cool as slants. (Slants can be obtained by placing the test tube rack at a 45° angle as the medium cools.)

Explant Preparation

Cotton callus was established in the "Explant Orientation" exercise (use cultures that are 4–6 weeks old).

Disinfestation of Explant Source

Callus is sterile.

Culture

1. Divide the callus into 23 equal-size portions (pea size).
2. Place 20 pieces of the callus on the nutrient medium. Place the cultures on the culture shelf.
3. Place the remaining 3 callus portions on a piece (~1 in.^{2}) of preweighed, labeled aluminum foil. Take a wet weight and place in the oven at 60°C; take a dry weight after 48 hr in the oven. Obtain averages for the initial wet and dry callus weights.

Tobacco

Purpose: To subculture and establish a growth curve for tobacco callus, *Nicotiana tabacum* L.

Explant Preparation

Callus induction has already been established. Tobacco callus was initiated by aseptically germinating seeds on a medium containing MS inorganic salts, 30 g/liter sucrose, and 8 g/liter TC agar at pH 5.7. In 3 weeks the seedlings were cut up and placed on a medium containing MS inorganic salts, 30 g/liter sucrose, 100 mg/liter myo-inositol, 0.5 mg/liter thiamine, 0.9 μM kinetin, 11.4 μM IAA, and 8 g/liter TC agar. The cultures were incubated in the dark at 27–30°C. In 6 weeks, callus was separated from the explant and subcultured for 6 weeks on the same medium for increase in callus volume.

Many experimental variations such as different concentrations of the five MS inorganic salt stocks, adding 0.3% (w/v) acid-washed charcoal to the medium, and varying the plant growth regulators and/or their concentrations can be tested.

Medium Preparation: 1 liter equivalent, Tobacco Growth Curve Medium.

1. Into a 2000-ml Erlenmeyer flask pour 500 ml of deionized, distilled water.
2. Mix in the following:
   a. 10 ml each Murashige and Skoog salts: nitrates, halides, NaFeEDTA, sulfates, and PBMo
   b. 10 ml thiamine stock (40 mg/liter)
   c. 10 ml myo-inositol stock (10 g/liter)
   d. 30 g sucrose
   e. 2 ml kinetin stock (10 mg/100 ml)
   f. 20 ml IAA stock (10 mg/100 ml)
3. Adjust volume to 1000 ml. Adjust pH to 5.7.
4. Add 8 g TC agar or Difco-Bacto agar. Melt.
5. Distribute 25 ml per culture tube (25 × 150 mm). Cap.
6. Autoclave for 15 min at 121°C, 15 psi. Cool as slants. (Slants can be obtained by placing the test tube rack at 45° angle as the medium cools.)

Disinfestation of Expant Source and Culture

Same as for "Cotton Growth Curve" exercise.

Observations

Before the callus is weighed, the agar can be gently blotted or wiped away with a paper towel so only callus weight is measured.

Data and Questions

1. Observe and take 0-, 5-, 10-, 20-, 25-, 30-, 35-, and 40-day samples. Take two samples at each point for dry and wet weights.
2. Establish a growth curve such as shown in Fig. 6.3.
3. What kind of statistical analysis is appropriate for these data?
3. Did all the explants grow equally? Why or why not?
4. How many parts per million of 2IP, IAA, NAA, kinetin, and 2,4-D are in the media, and what is the molarity of each?
5. Were there differences between the cotton and tobacco growth curves? If charcoal was used, was there a difference in the growth curve? Explain.
6. Did dry and wet callus weights correlate?

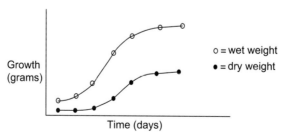

FIGURE 6.3   Take two samples every 5 days and plot wet and dry weights over time to establish growth curve.

CELLULAR VARIATION FROM CALLUS CULTURES

Plant cell cultures have the potential to be used for the production of valuable products by direct extraction from the cells or medium or biotransformation of a precursor into a desirable product. These products include alkaloids, steroids, amino acids, volatile oils, and saponins. Secondary plant compounds include cardiac stimulants, antitumor compounds, steroids, insecticides, and flavoring components. Traditionally, these compounds have been extracted from whole plants. However, this approach can be wasteful and labor intensive. Major potential advantages of using plant cell cultures are (a) the compounds can be produced under controlled environmental conditions; (b) the cultures are free of microbial and insect infestation; (c) the production of the compound can be manipulated and improved; and (d) with automation, the product production costs can be lower.

Callus cultures usually contain mixtures of cells that are not uniform in size, shape, pigmentation, metabolism, or chromosome number. Many differences are not visible. Some callus cultures give rise to cells of varying pigmentation which allows visual separation of the pigmented cell types and establishment of a uniform color type cell line. In some instances, these pigmented lines can be used in secondary metabolite isolation.

Purpose: Isolation and establishment of uniform pigmented cell lines from cotton or carrot explants.

Medium Preparation: Same as that for the "Explant Orientation" exercise or carrot callus induction.

Explant

Explant cultures are obtained from the "Explant Orientation" exercise on cotton or the "Carrot Callus Induction" exercise. After 6–8 weeks of incubation, callus should have differences in pigmentation. Visually separate the red, dark green, pale green, and yellow-beige cotton cell lines or the deep orange carrot cell lines and place callus on fresh medium. A similar study using cherry, peach, and Asiatic dayflower isolated cell lines high in anthocyanins, carotenoid pigmentation from Asiatic dayflower, betacyanin from *Amaranthus*, and isoflavone from soybean (Satoshi & Kazunori, 2011). Take final observations at 4 weeks.

Data and Questions

1. Do the cell lines grow at similar rates?
2. Are individual cells in the different cell lines identical in size and shape?
3. Do pigmentation changes occur over time in the different cell lines?

There are many interesting articles on secondary products resulting from cell cultures. For variations of the preceding exercise, refer to the references.

BIBLIOGRAPHY

Abe, T., & Futsuhara, Y. (1985). Efficient plant regeneration by somatic embryogenesis from root callus tissue of rice (*Oryza sativa* L.). *Journal of Plant Physiology, 121,* 111–118.

Alicchio, R., Antonioli, C., & Palenzona, D. (1984). Karyotypic variability in plants of *Solanum melongea* regenerated from callus grown in the presence of culture filtrate of *Verticillium dahliae*. *Theoretical and Applied Genetics, 67,* 267–271.

Bhaskaran, S., & Smith, R. H. (1988). Enhanced somatic embryogenesis in *Sorghum bicolor* from shoot tip culture. *In Vitro Cellular & Developmental Biology, 24,* 65–70.

Bhaskaran, S., & Smith, R. H. (1989). Control of morphogenesis in sorghum by 2,4-dichlorophenoxyacetic acid and cytokinins. *Annals of Botany, 64,* 217–224.

Bhaskaran, S., & Smith, R. H. (1990). Regeneration in cereal tissue culture: A review. *Crop Science, 30,* 1328–1337.

Binarova, P., Nedelnik, J., Fellner, M., & Nedbalkova, B. (1990). Selection for resistance to filtrates of *Fusarium* spp. in embryogenic cell suspension culture of *Medicago sativa* L. *Plant Cell Tissue & Organ Culture, 22,* 191–196.

Bohm, H. (1977). Secondary metabolism in cell cultures of higher plants and problems of differentiation. In M. Luckner, L. Nover, & H. Bohm (Eds.), *Secondary metabolism and cell differentiation* (pp. 105–123). New York: Springer-Verlag.

Close, K. R., & Gallagher-Ludeman, L. A. (1989). Structure–activity relationships of auxin-like plant growth regulators and genetic influences on the culture induction responses in maize (*Zea mays* L.). *Plant Science, 61,* 245–252.

Croughan, T. P., Stavarek, S. J., & Rains, D. W. (1981). *In vitro* development of salt resistant plants. *Environmental and Experimental Botany, 21,* 317–324.

D'Amato, F., Benniei, A., Cionini, P. G., Baroncelli, S., & Lupi, M. C. (1980). Nuclear fragmentation followed by mitosis as mechanisms for wide chromosome number variation in tissue cultures: Its implications for plantlet regeneration. In F. Sala, R. Parisi, R. Cella, & O. Ciferri (Eds.), *Plant cell cultures: Results and perspectives* (pp. 67–72). New York: Elsevier–North Holland.

DeFossard, R. A. (1974). Responses of callus from zygotal and micro-sporal tobacco (*Nicotiana tabacum* L.) to various combinations of indole acetic acid and kinetin. *New Phytology, 73,* 699–706.

Dhar, U., & Joshi, M. (2005). Efficient plant regeneration protocol through callus for *Saussurea obvallata* (DC.) Edgew. (Asteraceae): Effect of explant type, age and plant growth regulators. *Plant Cell Rep., 24,* 195–200.

Do, C. B., & Cormier, F. (1990). Accumulation of anthocyanins enhanced by a high osmotic potential in grape (*Vitis vinifera* L.) cell suspensions. *Plant Cell Reports, 9,* 143–146.

Evans, D. A., Sharp, W. R., & Flick, C. E. (1981). Growth and behavior of cell cultures: Embryogenesis and organogenesis. In T. A. Thorpe (Ed.), *Plant tissue culture: Methods and application in agriculture* (pp. 45–100). New York: Academic Press.

Fukui, H., Tani, M., & Tabata, M. (1990). Induction of shikonin biosynthesis by endogenous polysaccharides in *Lithospermum erythrorhizon* cell suspension cultures. *Plant Cell Reports, 9,* 73–76.

Furuya, T., Koge, K., & Orihara, Y. (1990). Long term culture and caffeine production of immobilized coffee (*Coffea arabica* L.) cells in polyurethane foam. *Plant Cell Reports, 9,* 125–128.

Gao, J., Li, J., Luo, C., Yin, L., Li, S., Yang, G., & He, G. (2010). Callus induction and plant regeneration in *Alternanthera philoxeroides*. *Molecular Biology Reports, 38*(2), 1413–1417.

Garcia, R., Pacheco, G., Falcao, E., Borges, G., & Mansur, E. (2011). Influence of type of explant, plant growth regulators, salt composition of basal medium, and light on callogenesis and regeneration in *Passiflora suberosa* L. (Passifloraceae). *Plant Cell Tissue and Organ Culture, 106,* 47–54.

Gifford, E. M., & Nitsch, J. P. (1969). Responses of tobacco pith nuclei to growth substances. *Planta, 85*, 1–10.

Gorham, J., Wyn Jones, R. G., & McDonnell, E. (1985). Some mechanisms of salt tolerance in crop plants. *Plant & Soil, 89*, 15–40.

Greenway, H., & Munns, R. (1980). Mechanisms of salt tolerance in nonhalophytes. *Annual Review of Plant Physiology, 31*, 149–190.

Hartman, C. L., McCoy, T. J., & Knous, T. R. (1984). Selection of alfalfa (*Medicago sativa*) cell lines and regeneration of plants resistant to the toxin(s) produced by *Fusarium oxysporum* F. sp. *medicaaginis*. *Plant Science Letters, 34*, 183–194.

Heinstein, P., & El-Shagi, H. (1981). Formation of gossypol by *Gossypium hirsutum* L. cell suspension cultures. *J. Nat. Prod., 44*, 1–6.

Helgeson, J. P., Kruger, S. M., & Upper, C. D. (1969). Control of logarithmic growth rates of tobacco callus tissue by cytokinins. *Plant Physiology, 44*, 193–198.

Heyser, J. W., & Nabors, M. W. (1982). Long-term plant regeneration, somatic embryogenesis and green spot formation in secondary oat (*Avena sativa*) callus. *Zeitschriftfuer Pflanzenphysiologie, 107*, 153–160.

Hodges, T. K., Kamo, K. K., Imbrie, C. W., & Becwar, M. R. (1986). Genotype specificity of somatic embryogenesis and regeneration in maize. *Biotechnology, 4*, 219–223.

Irvani, N., Salouki, M., Omidi, M., Zare, A. R., & Shahnuzi, S. (2010). Callus induction and plant regeneration in *Dorem ammoniacum* D., an endangered medicinal plant. *Plant Cell Tissue and Organ Culture, 100*, 293–299.

Kendall, E. J., Qureshi, J. A., Kartha, K. K., Leung, N., Chevrier, N., Caswell, K., & Chen, T. H. H. (1990). Regeneration of freezing-tolerant spring wheat (*Triticum aestivum* L.) plants from cryoselected callus. *Plant Physiology, 94*, 1756–1762.

Lee, S. L., & Scott, A. I. (1979). The industrial potentials of plant tissue culture. *Developments in Industrial Microbiology, 20*, 381–391.

Liu, K. C. S., Yang, S. L., Roberts, M. F., & Phillipson, J. D. (1990). Production of canthin-6-one alkaloids by cell suspension cultures of *Brucea javanica* L. Merr. *Plant Cell Reports, 9*, 261–263.

Ma, H., Gu, M., & Liang, G. H. (1987). Plant regeneration from cultured immature embryos of *Sorghum bicolor* L. Moench. *Theoretical and Applied Genetics, 73*, 389–394.

Mantell, S. H., & Smith, H. (1983). Cultural factors that influence secondary metabolite accumulations in plant cell and tissue cultures. *Seminar Series of the Society for Experimental Biology, 18*, 75–108.

Meredith, C. P. (1984). Selecting better crops from cultured cells. In J. P. Gustafson (Ed.), *Gene manipulation in plant improvement* (pp. 503–528). New York: Plenum.

Mok, M., Gabelman, W. H., & Skoog, F. (1976). Carotenoid synthesis in tissue cultures of *Daucas carota* L. *Journal of the American Society of Horticultural Science, 101*, 442–449.

Nabors, M. W., Gibbs, S. E., Bernstein, C. S., & Meis, M. E. (1980). NaCl-tolerant tobacco plants from culture cells. *Zeitschrift fuer Pflanzenphysiologie, 97*, 13–17.

Nabors, M. W., Heyser, J. W., Dykes, T. A., & DeMott, K. J. (1983). Long-duration, high-frequency plant regeneration from cereal tissue cultures. *Planta, 157*, 385–991.

Naor, V., Ziv, M., & Zahavi, T. (2011). The effect of the orientation of stem segments of grapevine (*Vitis vinifera*) cv. Chardonnay on callus development in vitro. *Plant Cell Tissue and Organ Culture, 106*, 353–358.

Peng, J., & Hodges, T. K. (1989). Genetic analysis of plant regeneration in rice (*Oryza sativa* L.). *In Vitro Cellular and Developmental Biology, 25*, 91–94.

Peterson, G., & Smith, R. H. (1991). Effect of abscisic acid and callus size of American and international rice varieties. *Plant Cell Reports, 10*, 35–38.

Price, H. F., Smith, R. H., & Grumbles, R. M. (1977). Callus culture of six species of cotton (*Gossypium* L.) on defined media. *Plant Science Letters, 10,* 115–119.

Rai, M. K., Kalia, R. W., Singh, R., Gangola, M. O., & Dhawan, A. K. (2011). Developing stress tolerant plants through in vitro selection—An overview of the recent progress. *Environmental & Experimental Botany, 71*(1), 89–98.

Reinert, J., & Yeoman, M. M. (Eds.), (1982). *Plant cell and tissue culture: A laboratory manual.* New York: Springer-Verlag.

Rines, H. W., & Luke, H. H. (1985). Selection and regeneration of toxin-insensitive plants from tissue cultures of oats (*Avena sativa*) susceptible to *Helminthosporium victoria. Theoretical and Applied Genetics, 71,* 16–21.

Satoshi, A., & Kazunori, O. (2011). Production of phytochemicals by using habituated and long-term cultured cells. *Plant Biotechnology, 28*(1), 51–62.

Schroeder, C. A., & Davis, L. H. (1962). Totipotency of cells from fruit pericarp tissue *in vitro. Science, 138,* 595–119.

Skirvin, R. M. (1978). Natural and induced variation in tissue culture. *Euphytica, 27,* 241–266.

Smith, M. K., & McComb, J. A. (1983). Selection for NaCl tolerance in cell cultures of *Medicato sativa* and recovery of plants from a NaCl-tolerant cell line. *Plant Cell Reports, 2,* 126–128.

Stavarek, S. J., & Rains, D. W. (1983). Mechanisms for salinity tolerance in plants. *Iowa State Journal of Research, 57,* 457–476.

Tal, M., & Shannon, M. C. (1983). The response to NaCl of excised, fully differentiated and differentiating tissues of cultivated tomato, *Lycopersicon esculentum,* and its wild relatives, *L. peruvianum and Solanum penillii. Physiologia Plantarum, 59,* 659–663.

Tomes, D. T., & Smith, O. S. (1985). The effect of parental genotype on initiation of embryonic callus from elite maize (*Zea mays* L.) germplasm. *Theoretical and Applied Genetics, 70,* 505–509.

Umetani, Y., Kodakari, E., Yamamura, T., Tanaka, S., & Tabata, M. (1990). Glucosylation of salicylic acid by suspension cultures of *Mallotus japonicus. Plant Cell Reports, 9,* 325–327.

Wang, P. J., & Huang, L. C. (1976). Beneficial effects of activated charcoal on plant tissue and organ culture. *In Vitro, 12,* 260–262.

Wang, X.-D., Nolan, K. E., Irwanto, R. R., Sheahan, M. B., & Rose, R. J. (2011). Ontogeny of embryogenic callus in Medicago truncatula: The fate of pluripotent and totipotent stem cells. *Ann. Bot., 107*(4), 599–609.

Wenzel, G., & Foroughi-Wehr, B. (1990). Progeny tests of barley, wheat, and potato regenerated from cell cultures after *in vitro* selection for disease resistance. *Theoretical and Applied Genetics, 80,* 359–365.

Witham, F. H. (1968). Effect of 2, 4-D on the cytokinin requirement of soybean cotyledon and tobacco stem pith callus tissues. *Plant Physiology, 43,* 1455.

Wernicke, W., & Brettell, R. I. S. (1982). Morphogenesis from cultured leaf tissue of *Sorghum bicolor*–culture initiation. *Protoplasma, 111,* 19–27.

Wysokinska, H., & Giang, N. T. X. (1990). Selection of penstemide and serrulatoloside producing clones in *Penstamon serrulatus* by small-aggregate cloning. *Plant Cell Reports, 9,* 378–381.

Yeo, A. R., & Flowers, T. J. (1983). Varietal differences in the toxicity of sodium ions in rice leaves. *Physiologia Plantarum, 59,* 189–195.

Regeneration and Morphogenesis

Effective regeneration of intact plants from explants placed in culture, or from callus cultures, is a vital sequence in the implementation of successful biotechnology approaches in plant improvement. Obtaining whole plants from explants, callus or single cells is central to obtaining genetically engineered plants, cell selection for unique traits, somaclonal variants, rapid clonal propagation, virus-free plants, haploid and/or polyploidy plants, embryo rescue, germplasm storage, etc.

The pathways followed in regeneration are varied. In theory all cells contain the genetic capability to direct their development into a complete plant; they are totipotent. However, not all cells seem to be able to express totipotency. Explants are composed of highly differentiated cells such as leaf, stem, root, and floral tissue. Individual flower petals of St John's Wort initated shoot organogenesis (Palmer & Keller, 2010). Immature embryos, plant meristems, and other

Plant Tissue Culture. Third Edition. DOI: 10.1016/B978-0-12-415920-4.00007-4

meristematic cells in the vascular system are undifferentiated. Explants are wounded in the isolation process to be placed in culture and are generally stimulated to divide and proliferate callus, a wound response; Sugiyama (1999) indicated that explant wounding triggers dedifferentiation in the explants. Callus cells are generally differentiated, unorganized cell masses and there are regions of cells in the callus that can revert to an undifferentiated or meristematic condition to regenerate organs.

Meristematic regions can form within a callus and are capable of shoot, somatic embryo, or root formation, adventitious organogenesis. This is considered indirect organogenesis because it has a callus intermediary stage. Direct organogenesis is the formation of meristematic regions directly from a cell of the explant. There is no intervening callus stage, and shoots, somatic embryos, roots, anthers, and flowers can directly arise from the explant. Individual explants can have shoots arise both by direct and indirect organogenesis (Mallon et al., 2011).

The early work by Skoog and Miller (1957) and others discussed in Chapter 1 established the basic concepts in organogenesis from callus cultures regarding auxin to kinetin ratios. High auxin to kinetin favored root formation, the opposite favored shoot formantion, and intermediate levels enhanced callus proliferation (Skoog and Miller, 1957). From these beginnings much progress has been made in basic plant physiology, molecular biology (Meng et al., 2010) and plant growth regulators (Werner et al. 2001); however, many plant cultivars are still recalcitrant in vitro. Iqbal et al. (2011) demonstrated an increase in embryogenic potential from leaf explants from plants that had previously regenerated in vitro. For a fuller discussion of these topics refer to textbooks in these areas and articles such as Magyar-Tabori et al. (2010), and references cited therein, and Nordstrom et al. (2004).

Databases in library searches are an invaluable source of current information on published literature on a multitude of plant species and the status of in vitro culture. This is a valuable starting point to initiate a cell culture study to successfully establish cultures. The following exercises will provide a foundation for studies on in vitro morphogenesis.

CONTROLLED MORPHOGENESIS: POTATO TUBERIZATION

Propagation of potato plants through meristem tip culture is a routine procedure for the recovery of virus-free plants and the production of certified virus-free plants (Wang & Hu, 1982; Zapata et al., 1995). In crops vegetatively propagated, virus infection of plants is a problem that can reduce production to noneconomical levels. The apical meristems of infected plants are generally free from or have a very low level of virus. In potato, meristem tips measuring 0.3 to 0.6 mm are excised from plants grown in a growth chamber and are placed on a suitable medium for shoot growth. When rooted, shoots are transferred to soil and kept in insect-proof greenhouses where they are further tested for the absence of the virus. The large-scale multiplication of plants and production of seed tubers must be done in isolated fields where the chance of reinfection is minimal. In many areas of the world, effective control of insects in such locations is almost impossible and insect-transmitted systemic diseases are unavoidable. Alternative strategies have been developed to reduce the problem.

A technique to deal with field reinfestation of virus-free plants is the *in vitro* production of tubers that can then be distributed to the producer with little to no contamination. This technique has been extensively studied (Tovar *et al.*, 1985; Hussey & Stacy, 1984). Results show the importance of a cytokinin in the medium and certain environmental requirements for tuberization (Mingo-Castro *et al.*, 1974).

Often in plant tissue culture experiments major concerns are optimum cell culture medium, inorganic salts, and plant growth regulator combinations and concentrations. The effects of different gelling agents are not routinely considered. Agar is a polysaccharide derived from extracts of seaweed. Gelrite is an agar substitute also produced by microbial fermentation. Gelrite forms a very clear gel.

The objective of this study is to compare the effects of two different gelling agents on the tuber formation of two potato *Solanum tubersosum* L. varieties, "Russet Burbank" and "Superior." Certainly, the concentration of carbohydrate and kinetin has significant effects on tuberization, and these parameters can also be examined.

Purpose: To examine the effects of gelling agents on potato tuberization *in vitro*.

Medium Preparation: 1-liter equivalent, Potato Tuberization Medium.

1. Into a 1000-ml Erlenmeyer flask pour 500 ml of deionized, distilled water.
2. Mix in the following:
   a. 10 ml each Murashige and Skoog salts: nitrates, halides, NaFeEDTA, sulfates, and PBMo
   b. 62.5 ml kinetin stock (40 mg/liter)
   c. 10 ml myo-inositol stock (10 g/liter)
   d. 60 g sucrose
3. Adjust volume to 1000 ml. Adjust pH to 5.7.
4. Divide into two parts, 500 ml each:
   a. Add 1 g Gelrite
   b. Add 4 g Difco-Bacto agar
5. Melt; distribute 25 ml/culture tube (25 × 150 mm). Cap using a color code for each gelling agent.
6. Autoclave for 15 min at 121°C, 15 psi.

Explant

Plants of the two potato cultivars "Russet Burbank" and "Superior" are grown under sterile conditions in Magenta GA-7 containers. They were initiated into culture by taking sprout cuttings from potatoes purchased at a produce store. The tubers were set on a window ledge to green up and initiate sprout growth at the eye. Sprouts ½-inch long were excised and surface sterilized by washing in soapy water, dipping in 95% alcohol, sterilizing in 20% chlorine bleach for 15 min, and rinsing three times in sterile water. The tissue burned by the chlorine bleach was removed, and the shoot tip was cultured. The shoots were cultured on MS salts, 100 mg/liter myoinositol, 20 g/liter sucrose, and 0.8% agar and placed on the culture shelf. The sprouts grew to form shoots in 6 weeks, and the shoots were used for the experiment or subcultured on the same medium to maintain the shoot cultures.

Culture shoot segments with two nodes on the medium: for tuber development, place the cultures in the dark at 19°C. Growth is fine at 22–23°C, but 19°C is better. As a variation, some of the cultures can be placed in the light.

Data and Questions

Decide on experimental parameters to measure (weight, length, diameter, etc.) on the initial explant and what measurements you will gather at the end of 4–6 weeks to determine what effect gelling agent and light conditions have on tuberization *in vitro*. Be prepared to discuss the following questions:

1. Did the two cultivars respond in the same way to the gelling agent and culture conditions?
2. Was there a significant difference in final tuber development among treatments?

SOMATIC EMBRYOGENESIS

Somatic embryogenesis is a process by which somatic (non-gametic) cells undergo differentiation to form a bipolar structure containing both root and shoot axes. These somatic embryos are similar to zygotic embryos and can mature and germinate (Steward *et al.*, 1958a, 1958b; Reinert, 1958; Kato & Takeuchi, 1963; McWilliam *et al.*, 1974; Nomura & Komamine, 1985). The process is illustrated in Fig. 7.1.

To induce somatic embryogenesis in carrot and many other species, callus must be placed on an auxin-containing medium. After a short "auxin pulse," the cells are placed on hormone-free medium for embryo development (Fujimura & Komamine, 1979; Borkird *et al.*, 1986). Carrot cell clusters 50–100 μm in diameter, which contain 10–20 small, highly cytoplasmic meristematic cells, have a high potential for embryonic development (Raghavan, 1985; Sung *et al.*, 1984). The process also requires a reduced form of nitrogen (glycine, glutamine, yeast extract, or ammonium). Nitrogen solely in the form of NO_3^- seldom gives rise to embryos. Multiplication of plants from cell cultures by somatic embryogenesis has the potential for the highest rates of plant

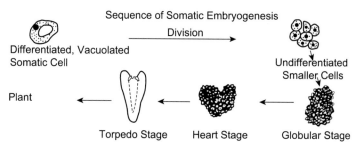

FIGURE 7.1   Through somatic embryogenesis a somatic (nongametic) cell undergoes differentiation to form a bipolar structure containing both root and shoot axes. This somatic embryo can mature and germinate.

production in culture. Thousands of somatic embryos can be produced in a single flask. However, normal maturation of all of these embryos into plants can be a problem. Seed companies have investigated this process for the production of "artificial seeds."

In this exercise, cells will be subcultured on an auxin-containing medium and placed on an auxin-free medium for embryo development.

Purpose: To induce somatic embryogenesis in suspension and callus cultures of carrot (*Daucus carota*) and observe the effect of inoculation density and 2,4-D on somatic embryogenesis.

Medium Preparation: 1 liter equivalent, Carrot Somatic Embryogeneis Medium.

1. Into a 1000-ml Erlenmeyer flask pour 500 ml of deionized, distilled water.
2. Mix in the following:
   a. 10 ml each Murashige and Skoog salts: nitrates, halides, NaFeEDTA, sulfates, and PBMo
   b. 10 ml thiamine stock (40 mg/liter)
   c. 10 ml myo-inositol stock (10 g/liter)
   d. 25 g sucrose
3. Adjust volume to 800 ml.
4. Place 400 ml in a 1000 ml flask labeled "A" and the remaining 400 ml in a 1000 ml flask labeled "B."
5. Add 0.5 ml 2,4-D stock (10 mg/100 ml) to the A flask.
6. Dilute both flasks to 500 ml. Adjust pH to 5.7.
7. Cap flasks with aluminum foil and autoclave 15 min at 121°C, 15 psi. Also autoclave a 100-ml beaker with four layers of cheesecloth covering the top, secured by several rubber bands or a string. Wrap the beaker in aluminum foil.
8. In a transfer hood dispense 20 ml into labeled (A and B) plastic Petri dishes (100 × 20 mm) or 10 ml into small plastic Petri dishes (60 × 15 mm).

Explant Preparation

Suspension cultures of carrot were grown on MS salts, 30 g/liter sucrose, 0.1 mg/liter thiamine, 0.5 mg/liter nicotinic acid, 0.5 mg/liter pyridoxine, 100 mg/liter myo-inositol, and 0.3 mg/liter 2,4-D. These were incubated on an orbital shaker (~90 rpm) at 27–30°C, under 240-fc, 24-h light. The original callus cultures were initiated in the "Aseptic Germination of Seeds" and "Callus Initiation" exercises.

Allow 2–3 months for callus induction and subculture to prepare for this exercise.

Procedure

1. Put the carrot suspension culture (~1 week in subculture) in the transfer hood and spray the outside with 70% alcohol. Allow to dry.
2. Aseptically pour the contents of the suspension culture onto the cheesecloth-covered beaker. This will deposit the carrot cells on the top of the cheesecloth. Now rinse the cells with fresh 2,4-D-free medium to wash the residual 2,4-D from the cells.
3. Use a sterile stainless steel chemical measuring scoop to inoculate the Petri dishes (A and B) with two or three different inoculation densities of cells. Generally a density of 10^3 to 10^5 clusters (50–100 μm in diameter) per milliliter of medium is a favorable inoculum for a high level of somatic embryo formation. Determine an average fresh weight for your inoculation densities to give you an idea of the fresh weight in milligrams. High yields of embryos have been observed at 20–50 mg fresh-weight callus per 20 ml of medium. Label your Petri dishes and seal with Parafilm.
4. Culture in the dark at 27–30°C on the orbital shaker.

Data and Questions

A dramatic effect of inoculation density on the incidence of somatic embryo formation will be observed. It is imperative that the cultures be examined under an inverted or dissecting microscope on a weekly basis for 1–4 weeks.

1. Draw the stages of embryogenesis and note when the different stages were first visible.
2. How many of these stages can you see in the A medium?
3. Are the embryos free floating or do they occur in clusters?
4. How would you attempt to establish these embryos as plantlets?
5. Does inoculation density have any affect on somatic embryogenesis?
6. What is the effect of 2,4-D?

Generally after 1 week globular embryos are apparent. At 1–3 weeks heart- and torpedo-shaped embryos should be visible.

REGENERATION OF RICE

Rice is an extremely important grain crop, and extensive studies on rice culture and organogenesis have been published. Genotype and the developmental stage of rice seeds are key factors in callus induction and regeneration potential in rice cultivars (Zuraida *et al.*, 2011; Lee *et al.*, 2002; Zuraida *et al.*, 2010). Mature rice seeds and mature and immature embryos are reported to be choice explants to obtain embryogenic cultures. Mature seeds as an explant source are convenient due to year-round availability; however, the mature seed has limits for recalcitrant genotypes. Zuraida *et al.* (2011) reviewed the process involved in optimizing explant medium and culture conditions to improve genotype response. A good study of the histology of rice somatic embryogenesis was published by Vega *et al.* (2009).

Purpose: To observe the regeneration potential of the two callus types of rice.

Medium Preparation: 1 liter equivalent, Rice Subculture Medium. The medium is the same as the medium for "Establishment of Competent Cereal Cultures" (Chapter 6) except add 26 mg/liter ABA and reduce 2,4-D to 2 mg/liter. Experimental variations can include testing different levels of ABA with and without different levels of 2,4-D.

Procedure: Prepare explants as in "Establishment of Competent Cereal Cell Cultures" using two callus types initiated from the explant. Culture both types of callus by using small 10-mg pieces. Culture in the dark. At the end of 3 weeks, the callus can be subcultured onto the same medium or moved to a plant regeneration medium.

Regeneration

Medium Preparation: 1 liter equivalent, Rice Plant Regeneration Medium. This medium is the same as the proceeding except eliminate ABA and 2,4-D. Add 0.5 mg/liter BA and 0.05 mg/liter NAA.

Procedure: Place the cultures in the dark for 1 week, then transfer to the culture shelf. In 4 weeks the plants will be ready for transfer to soil.

Data and Questions

1. What is the frequency of plant formation?
2. What function could ABA be performing in plant regeneration?
3. Did both callus types form plants?
4. What was the difference between the two rice cultivars in regard to plant formation?

DORMANCY REQUIREMENTS OF EXPLANTS

Purpose: To observe the effects of chilling on plant development from bulb scales.

Medium Preparation: 1 liter equivalent, Bulb Scale Medium.

1. Into a 2000-ml Erlenmeyer flask pour 500 ml of deionized, distilled water.
2. Mix in the following:
   a. 10 ml each Murashige and Skoog salts: nitrates, halides, NaFeEDTA, sulfates, and PBMo
   b. 10 ml thiamine stock (40 mg/liter)
   c. 10 ml myo-inositol stock (10 g/liter)
   d. 30 g sucrose
   e. 10 ml vitamin stock
   f. 100 ml BA stock (10 mg/100 ml)
   g. 10 ml NAA stock (10 mg/100 ml)
   h. 2 mg glycine
   i. 16 ml adenine sulfate stock (1 g/100 ml)
3. Adjust volume to 1000 ml. Adjust pH to 5.7.
4. Add 8 g TC agar or Difco-Bacto agar. Melt.
5. Distribute 25 ml/culture tube (25 × 150 mm). Cap.
6. Autoclave for 15 min at 121°C, 15 psi.
7. Cool as slants.

Explant Preparation

Remove bulb scales from narcissus or daffodil bulbs that have completed flowering (Seabrook *et al.*, 1976; Hussey, 1975). Compare explants from bulbs that have been treated for 6–8 weeks at 11°C to nontreated bulbs. Wash scales with soapy water and rinse 2 h in tap water. Surface sterilize in 95% ethanol for 10 s followed by 10% chlorine bleach for 20 min. Rinse three times in sterile water. Remove the distal and proximal regions of the bulb scales and cut sections of the immature leaf base 2 mm wide and 10 mm long. Insert the basal end into the agar (see Fig. 7.2). Place on the culture shelf. Experimental variations can include testing the effects of 1000 mg/liter case in hydrolysate in the medium and/or varying the BA and NAA in the medium.

Shoots can be transferred to the same medium as the preceding with 2 mg/liter BA and 2 mg/liter NAA for shoot development. Shoots can be rooted on a

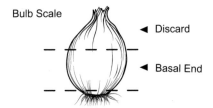

FIGURE 7.2   Plant the basal end of the bulb scale.

medium of half MS salts and half sucrose with no plant growth regulators. Sea-brook *et al.* (1976) reported obtaining 2620 plants from two leaf base explants; whereas with conventional methods, a maximum of 50 bulblets in 2 years had been reported.

Questions

1. How many shoots per explant developed?
2. Did the shoots originate from a certain area on the explant?
3. Was there any difference in the number of shoots formed between the cold-treated and non-cold-treated bulbs? Why?
4. If contamination was a problem, discuss possible sources of contamination.

PRIMARY MORPHOGENESIS: DOUGLAS FIR

Purpose: To observe direct morphogenesis from a primary explant of Douglas fir cotyledons.

Medium Preparation: 1 liter equivalent, Douglas Fir Morphogenesis Medium.

1. Into a 2000-ml Erlenmeyer flask pour 500 ml of deionized, distilled water.
2. Mix in the following:
   a. 10 ml each Murashige and Skoog salts: nitrates, halides, NaFeEDTA, sulfates, and PBMo
   b. 2.5 ml thiamine stock (40 mg/liter)
   c. 25 ml myo-inositol stock (10 g/liter)
   d. 30 g sucrose
   e. 11 ml BA stock (10 mg/100 ml)
   f. 0.01 ml NAA stock (10 mg/100 ml)
3. Adjust volume to 1000 ml. Adjust pH to 5.7.
4. Add 8 g TC or Difco-Bacto agar.
5. Autoclave for 15 min at 121°C, 15 psi.
6. Distribute 25 ml per sterile plastic Petri dish (100 × 20 mm).

Explant Preparation

Refer to the "Aseptic Germination of Seeds" exercise to obtain aseptically germinated Douglas fir seedlings. Contamination can still be a problem with aseptically germinated seeds. Excise the apical portion including cotyledons of a 2- to 4-week-old plant (Cheng, 1977, 1978; Sommer, 1975). Surface disinfect the seedling shoot in 10% chlorine bleach for 10 min; rinse three times in sterile water. Excise cotyledons from the axis of the seedling by severing them at the point of connection to the seedling. Slice cross sections of the cotyledons into 3- to 4-mm lengths. Discard the distal most 3 mm of the cotyledon. Culture the segments on the agar surface with three to five segments per dish. Incubate the cultures on the culture shelf.

Data and Questions

In 4 weeks observe the number of adventitious shoots.

1. Is there any callus formation?
2. What would be a major concern in establishing a forest with Douglas fir propagated in this fashion?
3. How many shoots developed from each explant?

BIBLIOGRAPHY

Ammirato, P. V. (1983). Embryogenesis. In D. A. Evans, R. Sharp, P. V. Ammirato, & Y. Yamada (Eds.), *Handbook of plant cell culture: Techniques for propagation and breeding* (Vol. 1, pp. 82–123). New York: MacMillan.

Bianchi, R., Fambrini, M., & Pugliesi, C. (1999). Morphogenesis in *Helianthus tuberosus*: Genotype influence and increased totipotency in previously regenerated plants. *Biologia Plantarum, 2*(4), 515–523.

Bonga, J. M. (1977). Applications of tissue culture in forestry. In J. Reinert, & Y. P. S. Bajaj (Eds.), *Applied and fundamental aspects of plant cell, tissue, and organ culture* (pp. 93–108). New York: Springer-Verlag.

Bonga, J. M. (1980). Plant propagation through tissue culture emphasizing wood species. In F. Sala, R. Parisi, R. Cella, & O. Ciferri (Eds.), *Plant cell cultures: Results and perspectives* (pp. 253–264). New York: Elsevier/North–Holland.

Borkird, C., Choi, J. H., & Sung, Z. R. (1986). Effects of 1, 2-dichloro-phenoxyacetic acid on the expression of embryogenic program in carrot. *Plant Physiology, 81,* 1143–1146.

Bourque, J. E., Miller, J. C., & Park, W. D. (1987). Use of an *in vitro* tuberization system to study tuber protein gene expression. *In Vitro Cellular & Developmental Biology, 23,* 381–386.

Cheng, T. Y. (1977). Factors affecting adventitious bud formation of cotyledon culture of Douglas fir. *Plant Science Letters, 9,* 179–187.

Cheng, T. Y. (1978). Clonal propagation of woody plant species through tissue culture techniques. *International Plant Properties Society, 28,* 139–155.

Durzan, D. J. (1980). *Progress and promise in forest genetics. Paper science technology—The cutting edge.* Appleton, WI: Institute of Paper Chemistry (pp. 31–60).

Fujimura, T., & Komamine, A. (1979). Involvement of endogenous auxin in somatic embryogenesis in carrot cell suspension culture. *Zeitschrift fuer Pflanzenphysiologie, 95,* 13–19.

Gautheret, R. J. (1966). Factors affecting differentiation of plant tissue *in vitro*. In W. Beerman (Ed.), *Cell differentiations and morphogenesis.* Amsterdam: North–Holland.

Halperin, W. (1969). Morphogenesis in cell cultures. *Annual Review of Plant Physiology, 20,* 395–418.

Hicks, G. S. (1980). Patterns of organ development in plant tissue culture and the problem of organ determination. *Botany Review, 46,* 1–23.

Hussey, G. (1975). Totipotency in tissue explants and callus of some members of the Liliaceae, Iridaceae and Amaryllidaceae. *Journal of Experimental Botany, 26*(91), 253–262.

Hussey, G., & Stacey, N. J. (1984). Factors affecting the formation of *in vitro* tubers of potato (*Solanum tuberosum* L.). *Annals of Botany, 53,* 565–578.

Iqbal, M. M., Nazir, F., Iqbal, J., Tehrim, S., & Zafar, Y. (2011). In vitro micropropagation of peanut (Arachis hypogaea) through direct somatic embryogenesis and callus culture. *Int. J. Agric. Biol., 13*, 811–814.

Kato, H., & Takeuchi, M. (1963). Morphogenesis *in vitro* from single cells of carrot root. *Plant & Cell Physiology, 4*, 243–245.

Konar, R. N., & Naagmani, R. (1974). Tissue culture as a method for vegetative propagation of forest trees. *New Zealand Journal of Forest Science, 4*, 279–290.

Krikorian, A. D., Dutcher, F. R., Quinn, C. E., & Steward, F. C. (1981). Growth and development of cultured carrot cells and embryos under space flight conditions. In W. R. Holmquist (Ed.), *Advances in space research (COSPAR, 1980)* (Vol. 11, pp. 117–127). New York: Pergamon.

Lee, K. S., Jean, H. S., & Kim, M. Y. (2002). Optimization of a mature embryos-based in vitro culture system for high-frequency somatic embryogenenic callus induction and plant regeneration from japonica rice cultivar. *Plant Cell, Tissue and Organ Culture, 71*, 9–13.

Magyar-Tabori, K., Dobranszki, J., Teixeira da Silva, J. A., Bulley, S. M., & Hudak, I. (2010). The role of cytokinins in shoot organogenesis in apple. *Plant Cell Tiss Organ Cult, 101*, 31–39.

Mallon, R., Rodriguez-Oubina, J., & Luz Gonzalez, M. (2011). Shoot regeneration from *in vitro*-derived leaf and root explants of *Centaurea ultreiae*. *Plant Cell Tissue and Organ Culture, 106*(3), 523–530.

McWilliam, A. A., Smith, S. M., & Street, H. E. (1974). The origin and development of embryoids in suspension cultures of carrot (*Daucus carota*). *Annals of Botany, 38*, 243–250.

Meng, L., Zhanag, S., & Lemaux, P. G. (2010). Toward molecular understanding of in vitro and in planta shoot organogenesis. *Critical Reviews in Plant Sciences, 29*(2), 108–122.

Mingo-Castel, A. M., Negm, F. B., & Smith, O. E. (1974). Effect of carbon dioxide and ethylene on tuberization of isolated potato stolons cultured *in vitro*. *Plant Physiology, 53*, 798–801.

Mott, R. L. (1978). Tissue culture propagation of conifers. In K. Hughes, R. Henke, & M. Constatin (Eds.), *Propagation of higher plants through tissue culture* (pp. 125–133). Washington DC: US Technical Information Center.

Nomura, K., & Komamine, A. (1985). Identification and isolation of single cells that produce somatic embryos at a high frequency in carrot suspension culture. *Plant Physiology, 79*, 988–991.

Nordstrom, W., Tarkowski, P., Tarkowski, D., Norbaek, R., Astot, C., Dolezal, K., & Sandberg, G. (2004). Auxin regulation of cytokinin biosynthesis in *Arabidopsis thaliana*: A factor of potential importance for auxin-cytokinin-regulated development. *Proc. Natl. Acad. Sci. USA, 101*, 8039–8044.

Palmer, C. D., & Keller, W. A. (2010). Plant regeneration from petal explants of *Hypericum perforatum* L. *Plant Cell Tissue & Organ Culture, 104*(1), 91–100.

Raghavan, V. (1985). *Embryogenesis in angiosperms: A developmental and experimental study.* New York: Cambridge Univ. Press.

Reinert, J. (1958). Morphogenese und ihre Kontrolle an Gewebekulturen aus Karotten. *Naturwissenschaften, 45*, 344–345.

Seabrook, J. E. A., Cumming, B. G., & Dionne, L. H. (1976). *In vitro* induction of adventitious shoot and root apices on *Narcissus* (daffodil and narcissus) cultivar tissue. *Canadian Journal of Botany, 54*, 814–819.

Seabrook, J. E. A., & Cumming, B. G. (1977). The *in vitro* propagation of amaryllis (*Hippeastrum* spp Hybrids). *In Vitro, 13*(12), 831–836.

Skoog, F., & Miller, C. O. (1957). Chemical regulation of growth and organ formation in plant tissue cultures *in vitro*. *Symposium of the Society of Experimental Biology, 11*, 118–131.

Skoog, F., & Miller, C. O. (1957). Chemical regulation of growth and organ formation in plant tissue cultured *in vitro*. *Symposia of the Society for Experimental Biology, 11*, 118–130.

Smith, S. M., & Street, H. E. (1974). The decline of embryogenic potential as callus and suspension cultures of carrot (*Daucas carota* L.) are serially subcultured. *Annals of Botany, 38*, 223–241.

Sommer, H. E. (1975). Differentiation of adventitious buds on Douglas fir embryos *in vitro*. *Proceedings of the International Plant Properties Society, 25*, 125–127.

Sondahl, M. R., Caldas, L. S., Maraffa, S. B., & Sharp, W. R. (1980). The physiology of asexual embryogenesis. *Horticulture Review, 2*, 268–310.

Steward, F. C., Mapes, M. O., & Mears, K. (1958a). Growth and organized development of cultured cells. I. Growth and division of freely suspended cells. *American Journal of Botany, 45*, 693–703.

Steward, F. C., Mapes, M. O., & Mears, K. (1958b). Growth and organized development of cultured cells. II. Organization in cultures grown from freely suspended cells. *American Journal of Botany, 45*, 705–708.

Sugiyama, M. (1999). Organogenesis in vitro. *Curr. Opin. Plant Biol., 2*, 61–64.

Sung, Z. R., Feinberg, A., Chorneau, R., Borkird, C., Furner, I., Smith, J., Terzi, M., LaSchiavo, R., Giuliano, G., Pitto, L., & Nutti-Ronchi, V. (1984). Developmental biology of embryogenesis from carrot culture. *Plant Molecular Biology Reports, 2*, 3–14.

Tovar, P., Estrada, R., Schilde-Pentschler, L., & Dodds, J. H. (1985). Induction and use of *in vitro* potato tubers. *International Potato Center, 13*, 1–5.

Vega, R., Vasquez, N., Espinoza, A. M., Gatica, A. M., & Melara, M. V. (2009). History of somatic embryogenesis in rice (Oryza sativa CU. 5272). *Rev Biol. Trop., 57*(i), 144–150.

Wang, P., & Hu, C. (1982). *In vitro* mass tuberization and virus-free seed potato production. *Taiwan American Potato Journal, 59*, 33–37.

Warren, G. S., & Fowler, M. W. (1977). A physical method for the separation of various stages in the embryogenesis of carrot cell cultures. *Plant Science Letters, 9*, 71–76.

Werner, T., Motyka, V., Strnad, M., & Schmulling, T. (2001). Regulation of plant growth by cytokinin. *PNAS, 98*(18), 10487–10492.

Westcott, R. J., Henshaw, G. G., & Roca, W. A. (1977). Tissue culture storage of potato germplasm: Culture initiation and plant regeneration. *Plant Science Letters, 9*, 309–315.

Zapata, C. C., Miller, J. C., & Smith, R. H. (1995). An *in vitro* procedure to eradicate potato viruses X, Y, and S from Russet Norkotah and two of its strains. *In Vitro Cellular & Developmental Biology, Plant, 31*, 153–159.

Zuraida, A. R., Sur, R., WanZaliha, W. S., & Sreeramanan, S. (2010). Regeneration of Malaysian indica rice (*Oryza sativa* L.) variety MR 232 via optimized somatic embryogenesis system. *J. Phytol., 2*(3), 30–38.

Zuraida, A. R., Naziah, B., Zamri, Z., & Sreeramanan, S. (2011). Efficient plant regeneration of Malaysian indica rice MR219 and 232 via somatic embryogenesis system. *Acta Physiologiae Plantarum, 33*(5), 1913–1921.

Chapter 8

Woody Shrubs and Trees

Brent H. McCown
University of Wisconsin

Miroculture has been used with increasing frequency to manipulate ornamental shrubs and trees; however, the use of this technique for woody perennial crops can often be more complex than parallel applications for herbaceous crops. The reason for this difference is generally due to the relative difficulty of microculturing woody perennials as compared to herbaceous plants. Many of these problems can be placed under the term "recalcitrance" and can be traced to a number of factors, principal among which is the complex vegetative life cycle and generally slow growth of woody plants (McCown, 2000). Strong seasonal bud dormancy, predetermined and limited periods of seasonal shoot growth, and a marked change in growth characteristics as the plants progress from the juvenile to adult phases of their life cycle all complicate responsiveness and predictability in microculture. For example, trees and shrubs can be successfully microcultured at commercial scales only in the juvenile (rejuvenated) phase of the life cycle; the same plant when in the adult phase of its life cycle will be difficult or even impossible to establish in microculture. Thus, seedling sources of tissue must be used for either establishment in culture or a long period of "rejuvenation," either by manipulation of stock plants or by careful sequential *in vitro* subculturing of growing buds, and must be conducted before successful microcultures can be established. The physiology behind such recalcitrance is not fully understood, but views as to its

Plant Tissue Culture. Third Edition. DOI: 10.1016/B978-0-12-415920-4.00008-6

biological basis and methods to limit its impact can be found in a number of reviews (McCown, 1986; McCown & McCown, 1986; McCown, 2000.)

The growth rates of shoots and other tissues such as callus of woody perennials in microculture are often slower than observed with herbaceous plants. This slow growth can complicate all stages of microculture. Contamination of isolated tissues is more difficult to overcome because the contaminating organism can easily outgrow and overwhelm the slowly responding plant tissues. The production of microshoots may require months of growth instead of weeks, making the economics of *in vitro* cloning less tenable. Regeneration of adventitious organs is also relatively slow, thus complicating shoot recovery from genetic engineering experiments and increasing both the cost and risk of micropropagule production.

Most woody crops are first established in microculture using *shoot culture* approaches where the stimulation of axillary buds is the prime objective. Once a vigorous shoot culture is achieved, then the shoot culture itself becomes a convenient stock providing tissues for biotechnological manipulations or microshoots for cloning (McCown, 1985). This reliance on shoot cultures as a general source of tissues has the distinct advantage of eliminating the problems associated with seasonal growth, dormancy, and long *in vitro* establishment times that have to be faced every time one reisolates new explants of woody species for any application.

One prominent aspect of woody perenial shoot culture is "stabilization." Stabilization refers to a period when shoots are actively maintained in microculture and during which the growth characteristics of the shoots change to a type of growth more acclimated to the microculture environment. At the beginning of stabilization (after the initial successful isolation of buds or shoots), the earliest shoot growth is typified by irregular growth flushes that often produce abnormally appearing shoots with large, distorted leaves. Excess callus production is also often observed at this early stage of stabilization. As any new shoots produced in microculture are subcultured on a regular basis, the shoot growth appearance progressively changes to one typified by reproducible and continuous growth of uniform shoots with small leaves; ideally little callusing is evident. The most obvious indication that the stabilization phase is complete is that each subculture yields the same type of shoot culture (uniform and continuous growth rate and uniform appearance of the shoots) as was evident in the previous subculture. Microshoots harvested from such stabilized shoot cultures respond more uniformly and successfully to rooting and acclimation than do microshoots harvested from unstabilized stock shoot cultures. Likewise, tissues, cells, and protoplasts isolated from stabilized shoot cultures are more responsive (Russell & McCown, 1986). The period required for stabilization to be completed can vary widely depending on the plant, its history, and other factors. Some cultures can be stabilized after two to five subcultures. Other cultures have required more than 20 subcultures (more than 2 years); some plants (such as most species of *Quercus*) have never been successfully stabilized as shoot cultures. Thus, the application of biotechnology and micropropagation to some woody plant crops has been limited.

Even with these obvious disadvantages, microculture has been successfully applied to a wide range of trees and shrubs (Jain, 2003). This is particularly true with the application of micropropagation technologies. Successful species have a number of similar characteristics. Biologically, the successful crops are usually the faster growing trees and shrubs. Economically, only those selections that can claim a relatively high value in the market have been commercialized successfully. High value is critical since microcultured propagules of trees and shrubs are considerably more expensive than herbaceous crop counterparts due to increased production times in microculture.

The mineral salt medium most widely used for the microculture of woody crops is Woody Plant Medium (WPM) (Lloyd & McCown, 1980) shown in Table 8.1. The formulation was developed after it was discovered that many woody plants could not tolerate the relatively high salt, high chloride levels encountered in medium formulations like MS.

The widest application of microculture to trees and shrubs has been micropropagation. The following exercises are based on successful micropropagation protocols used in commercial applications. Two major problems that can complicate the use of woody perennials in an academic setting should be noted:

1. Obtaining suitable plant material for isolation may be difficult. Optimally, the material should be in active growth and obtained from seeds, seedlings, or newly propagated plants. The examples below were chosen in part because material should be available year-round in most localities.

2. The tissues may be slow to respond in culture. Thus if it is desired to show all stages of the microculture process, then it may be appropriate to start different student groups at different steps in the micropropagation process. For example, some groups can do isolation, some groups subculturing and optimization using already established shoot cultures, and other groups rooting and acclimation of microshoots.

RHODODENDRONS AND AZALEAS

Members of the Ericaceae were some of the earliest woody shrubs to be successively micropropagated on a commercial scale (McCown, 1986; McCown & McCown, 1986; Eeckhaut et al., 2010; Economou & Read, 1984; Ettinger & Preece, 1985; McCowan & Lloyd, 1983). Kalmia, Rhododendron, and Vaccinium are now all widely micropropagated; some of these crops can only be cloned commercially through micropropagation. One reason for the relative success with micropropagation is the ease with which these plants can be stabilized in culture. Ericaceae as a group are intolerant of BA and MS, both of which produce stunted shoot growth. For that reason, 2iP or Zeatin in WPM is commonly used.

Purpose: To propagate Rhododendron sp. by axillary bud proliferation (shoot cultures).

Medium Preparation: WPM with 1 μM 2iP (Table 8.1).

TABLE 8.1 Woody Plant Medium[a]

Stock	Constituents	g/liter[b]	ml/liter[c]	mg/liter[d]
A	NH_4NO_3	20.0	20	400
	$Ca(NO_3)_2 \cdot 4H_2O$	27.8		556
B	K_2SO_4	49.5	20	990
C	$CaCl_2 \cdot 2H_2O$	19.2	5	96
D	KH_2PO_4	34.0	5	170
	H_3BO_3	1.24		6.2
	$Na_2MoO_4 \cdot 2H_2O$	0.05		0.25
E	$MgSO_4 \cdot 7H_2O$	74.0	5	370
	$MnSO_4 \cdot H_2O$	3.38		16.9
	$ZnSO_4 \cdot 7H_2O$	1.72		8.6
	$CuSO_4 \cdot 5H_2O$	0.005		0.025
F[e]	$FeSO_4 \cdot 7H_2O$	5.57	5	27.8
	Na_2EDTA	7.45		37.3
G[f]	Thiamine · HCL	0.2	5	1.0
	Nicotinic acid	0.1		0.5
	Pyridoxine · HCL	0.1		0.5
	Glycine	0.4		2.0
H	Myo-inositol	20.0	5	100

Other constituents	g/liter
Sucrose	20
Calcium gluconate (Sigma #G-4625)[g]	1.3
Agar (Sigma A-1296)	3.0
Gelrite (Scott Labs 4900-1891)	1.1

Note. All stocks (but G) are made in bulk and autoclaved to insure sterility. The pH of the medium is usually adjusted to the desired level (commonly 5.6) using KOH after adding the gelling agents.
[a]*Lloyd, G. B., & McCown, B. H. (1980). International Plant Propagation Society, 30, 421–427.*
[b]*Stock concentration.*
[c]*Volume of stock to add to medium.*
[d]*Final concentration in medium.*
[e]*This stock is prepared by dissolving each component separately in half the final volume of distilled water (heat required) and then combining the two solutions to make a yellow stock.*
[f]*This stock is filter sterilized and kept frozen in 10- to 20-ml aliquots.*
[g]*Zeldin, E. L., & McCown, B. H. (1986). Abstracts of the VIth IAPTC Congress, p. 57.*

Explants

1. Obtain an actively growing plant or shoots of an azalea or broad-leaved rhododendron. These can either be from a garden or from a potted plant (greenhouse or garden center). Ideally, select shoots that originate from the root/shoot interface (collar) region, as these shoots will be the most juvenile. Isolate the shoot-tip (top 1 inch) and several 1- to 2-inch sections below this tip.
2. Remove all leaves that can be easily cut from the stem.
3. Surface-sterilize (with vigorous shaking) for 10 to 15 min in 10% (v/v) chlorine bleach with 2 drops/100 ml of Tween-20 and rinse three times in sterile water.
4. Cut the stem into individual stem pieces each containing a shoot tip or one to three nodes.
5. Place each explant basal-end-down into the medium to a depth of about $\frac{1}{4}$ of its length.
6. Place under long days (16 to 24 h light).
7. As shoots emerge from the preformed buds on these explants, remove them as soon as they reach about 1 cm in length and place them on fresh medium. Removal of the tips on these shoots will increase branching. Repeat this procedure as rapidly as possible and observe the change in shoot quality and responsiveness (stabilization) as the number of subcultures increases.

Advanced Studies

Once actively growing shoot cultures have been generated, experiments demonstrating the effect of a 2iP response curve (0, 0.1, 1, 10, and 40 μM) can be conducted by subculturing shoots onto each of these media.

Microshoots no less than 2 cm in length can be harvested, placed in water, and then stuck into a clean rooting medium (e.g., 1:1 peat/vermiculite or peat/sand) in a closed humid clear-plastic box. Place under indirect light at room temperature and mist the cuttings as needed (but do not saturate the medium). Rooting should be observed within a month.

Other Ericaceous crops can also be experimented with, including blueberries.

ROSES

In general, micropropagation is not used extensively for roses as they can be readily cloned by conventional cuttings or by grafting. However, micropropagation is used to rapidly increase the stock of new selections (Dubois *et al.*, 1988; Hasegawa, 1979; Pati *et al.*, 2005). These micropropagated stock plants can then be used in conventional propagation. Also, micropropagated plants have been grown in small pots and flowered early to create some unique flowering pot plants, which after their use indoors can be planted outdoors as ornamentals.

Purpose: To propagate *Rosa* sp. by axillary bud proliferation (shoot cultures).

Medium Preparation: MS inorganic salts, 30 g/liter sucrose, 100 mg/liter myo-inositol, MS vitamins, 8 g/liter TC agar with 1 μM BA.

Explant

1. Obtain actively growing shoots. These can be from the garden, a greenhouse, or a potted plant.
2. Proceed with these shoots as described previously for rhododendron.

Advanced Studies

Once actively growing shoot cultures have been generated, experiments demonstrating the effect of a BA response curve (0, 1, 4, and 10 μM) can be conducted by subculturing shoots onto each of these media.

Microshoots no less than 2 cm in length can be harvested, placed in water, and then stuck into a clean rooting medium (e.g., 1:1 peat/vermiculite or peat/sand) in a closed humid clear-plastic box. Place under indirect light at room temperature and mist the cuttings as needed (but do not saturate the medium). Rooting should be observed within a month.

BIRCH TREES

Ornamental white-barked birches were one of the first trees to be micropropagated commercially (Magnusson *et al.*, 2009; McCown, 1986 & 1989; McCown & Amos, 1979). Although trees often require longer times for stabilization, once this has been achieved, the faithfully subcultured shoot cultures can remain productive indefinitely. One birch cultivar has been in commercial production as a shoot culture for more than three decades.

Purpose: To propagate *Betula* sp. by axillary bud proliferation (shoot cultures).

Medium Preparation: WPM with 4 μM BA.

Explant

Two general approaches can be taken to establish birch in microculture:

1. *Shoots from established plants:* Seedlings or young plants will respond the most rapidly. Ideally, select shoots that originate from the root/shoot interface (collar) region, as these shoots will be the most juvenile. Obtain actively growing shoots from a young birch tree. If these are not available, obtain nongrowing but nondormant (mid- to late winter) 1- to 2-ft. branches of a young birch tree. Place these to a depth of 4 to 6 inches in clean water that is constantly well aerated; use an area with 18 to 24 h lighting and room temperatures. Within several weeks, the buds should be forced into new growth and will form small shoots.

2. *In vitro germination of birch seed.* Unfortunately, birch seed is not long lived and thus it may not be possible to obtain viable seed year-round. A general method follows:
   a. Place 150 mg of seed into a net bag constructed out of window screening or pantyhose. Seal open end with a paper clip.
   b. Treat for 30 s in 70% ETOH.
   c. Treat for 30 min in 10% bleach.
   d. Rinse in sterile water.
   e. Remove seed from the bag and place seed on medium (MS inorganic salts without hormones) surface. Germination usually occurs in 4 to 7 days and usable shoots can be obtained in several weeks.

Proceed with shoots isolated by any of the above procedures by following the procedures as described previously for rhododendron. Obviously, the shoots from the *in vitro* germinated plants do not need to be resterilized.

Advanced Studies

Once actively growing shoot cultures have been generated, experiments demonstrating the effect of a BA response curve (0, 1, 4, and 10 μM) can be conducted by subculturing shoots onto each of these media.

Microshoots no less than 2 cm in length can be harvested, placed in water, and then stuck into a clean rooting medium (e.g., 1:1 peat/vermiculite or peat/sand) in a closed humid clear-plastic box. Place under indirect light at room temperature and mist the cuttings as needed (but do not saturate the medium). Rooting should be observed within a month.

WHITE CEDAR (ARBORVITAE)

Gymnosperms are rarely easy to isolate and grow successfully in microculture. *Thuja* is an exception, although the isolates may take some time before active shoot growth is observed. *Thuja* culture is very instructive, however, because the shoot cultures will visibly change with progressive subcultures from the adult, flat-leaved form to the juvenile, needle-leaved form of growth.

Purpose: To propagate a gymnosperm by axillary bud stimulation (shoot cultures).

Medium Preparation: WPM with 1 μM BA.

Explant

1. Obtain 2- to 3-inch long sections of new season growth. Again, the younger the stock plant, the more likelihood of success. Potted stock from a nursery is often ideal as this can be forced into growth anytime of the year.
2. Cut the tips of the newest growth into 0.5- to 1-inch sections.

3. Surface-sterilize (with vigorous shaking) for 15 to 20 min in 10% (v/v) chlorine bleach with 2 drops/100 ml of Tween-20 and rinse three times in sterile water.
4. Place each explant basal-end-down into the medium to a depth of about ¼ of its length.
5. Place under long days (16 to 24 h light).
6. As shoots emerge from these preformed buds on these explants, remove them as soon as they reach 0.5 cm or more in length and place them on fresh medium. Repeat this procedure as rapidly as possible and observe the change in appearance of the shoots and responsiveness (stabilization) as the number of subcultures increases.

REFERENCES

Dubois, L. A. M., Roggemans, J., Soyeurt, G., & Devries, D. P. (1988). Comparison of the growth and development of dwarf rose cultivars propagated *in vitro* and *in vivo* by softwood cuttings. *Scientia Horticulturae, 35*, 293–299.

Economou, A. S., & Read, P. E. (1984). *In vitro* shoot proliferation of Minnesota deciduous azaleas. *HortScience, 19*, 60–61.

Eeckhaut, T., Janssens, K., De Keyser, E., & De Reik, J. (2010). Micropropagation of Rhododendron. *J. Methods Mol. Biol., 589*, 141–152.

Ettinger, T. L., & Preece, J. E. (1985). Aseptic micropropagation of Rhododendron PJM hybrids. *Journal of Horticultural Science, 60*, 269–274.

Hasegawa, P. M. (1979). *In vitro* propagation of rose. *HortScience, 14*, 610–612.

Jain, S. M., & Ishii, K. (Eds.), (2003). *Micropropagation of woody trees and fruits. springer Forestry Sciences* (Vol. 75). p. 852 ISBN 978-1-4020-1135-1.

Lloyd, G., & McCown, B. (1980). Commercially feasible micropropagation of mountain laurel, *Kalmia latifolia*, by use of shoot-tip culture. *Combined Proceedings of the International Plant Properties Society, 30*, 421–427.

McCown, B. H. (1985). From gene manipulation to forest establishment: Shoot cultures of woody plants can be a central tool. *TAPPI J., 68*, 116–119.

McCown, B. H. (1986). Woody ornamentals, shade trees, and conifers. In R. H. Zimmerman, R. J. Griesbach, F. A. Hammerschlag, & R. H. Lawson (Eds.), *Tissue culture as a plant production system for horticultural crops* (pp. 333–342). Dordrecht, The Netherlands: Martinus Nijhoff.

McCown, B. H. (1989). *Betula*, the birches. In Y. P. S. Bajaj (Ed.), *Biotechnology in agriculture and forestry: Trees II* (Vol. 5, pp. 324–341). New York: Springer-Verlag.

McCown, B. H. (2000). Recalcitrance of woody and herbaceous perennial plants: Dealing with genetic predeterminism. *In vitro Cell. Develop. Biol. Plant, 36*, 149–154.

McCown, B. H., & Amos, R. (1979). Initial trials with commercial micropropagation of birch selections. *Combined Proceedings of the International Plant Properties Society, 29*, 387–393.

McCown, B. H., & Lloyd, G. B. (1983). A survey of the response of *Rhododendron* to *in vitro* culture. *Plant Cell Tissue & Organ Culture, 2*, 77–85.

McCown, D. D., & McCown, B. H. (1986). North American hardwoods. In J. M. Bonga, & D. J. Durzan (Eds.), 2nd ed. *Cell and tissue culture in forests. Case histories: Gymnosperms, angiosperms and palms* (Vol. 3, pp. 247–260). Dordrecht, The Netherlands: Martinus Nijhoff.

Magnusson, V. A., Castillo, C. M., & Dai, W. (2009). Micropropagation of two elite birch species through shoot proliferation and regeneration. *Acta Hort. (ISHS), 812,* 223–230. http://www. actahort.org/books/812/812_28.htm.

Pati, P. K., Rath, S. P., Sharma, M., Sood, A., & Ahuja, P. S. (2005). *In vitro propagation of rose—a review.* Available on line at www.sciencedirect.com. (Biotechnology Advances, Elsevier), pp. 95–109.

Russell, J. A., & McCown, B. H. (1986). Culture and regeneration of *Populus* leaf protoplasts isolated from non-seedling tissue. *Plant Science, 46,* 133–142.

Haploid Plants from Anther Culture

The purpose of anther and pollen culture is the production of haploid plants through the induction of androgenesis (development of haploids from the male gamete) in the haploid cells of the immature pollen grain. Descriptions of microspore developmental stages using light and electron microscopy are described by Seguí-Simarro and Nuez (2007) and Raghavan (1997), as well as in plant anatomy texts.

Haploid plants are important for a number of reasons. Because they possess only a single set of chromosomes, even recessive mutations are phenotypically expressed. Plant breeders are especially interested in haploid plants because either spontaneous doubling of the chromosome number (to *2N*) by endomitosis or an application of the chemical colchicine or other antimitotic agents (Dhooghe *et al.*, 2011) to double the chromosome number can give rise to homozygous plants. These plants can be selected for desirable characteristics and used as hybrid parents without the normal three to five generations needed to produce stable homozygous lines. Additionally, doubled haploids are useful for molecular mapping (Croser *et al.*, 2006).

Multiple factors influence the success of anther culture. The physical status of the donor plant, inluding nutritional status and physiological age, environmental conditions like temperature and photoperiod, and the genotype (although

Plant Tissue Culture. Third Edition. DOI: 10.1016/B978-0-12-415920-4.00009-8

some genotypes will not respond) are highly significant. Generally, anthers from flowers produced early are more responsive. The developmental stage of the pollen is very important; the uninucleate stage is usually most responsive. Seguì-Simarro and Nuez (2007) describe in detail their approach to identify microspore developmental stage with the flower bud size. There are also a number of pretreatments of either the flower bud (generally 5°C for several hours to several days) or the entire plant that are essential in some species.

Dunwell (1976) showed that both the light intensity and photoperiod under which the donor plant is grown, as well as the temperature and nutrient status of both the plant and culture medium, can affect the yield of haploid plants from anther culture. Donor plants grown in the greenhouse obviously are easier to treat than much larger plants like trees and field-grown plant material. The composition of the nutrient medium, including inorganic salt formulations, carbohydrate source, the inclusion of activated charcoal, plant growth regulator concentrations and combinations, as well as the medium solidifying agent, are sometimes critical. Sometimes subculture and medium changes are required, as well as changes in the physical culture condition of temperature, light intensity and photoperiod. Additionally, there are the challenges of the very low response of cultured anthers, sometimes requiring thousands of anthers for each treatment, production of mixiploids, and the very real problem of albino plants that result from anther culture.

Given all these parameters to optimize, the process of having success generally starts with a literature review. There are many excellent published papers reporting successful approaches (Gosal et al., 2010; Ferrie & Mollers, 2011; Sundar & Jawahar, 2010; Croser et al., 2006; Ferrie et al., 2011; Germanà, 2011a,b; Dunwell, 2010; Kim & Baeziger, 2005).

As plantlets emerge from the anther, they can be haploid, diploid or mixoploid, and in many cases albino. In sweet pepper, Luitel and Won Hee (2011) found haploid plants were short, narrow-leaved, and had smaller fruit than diploid plants. Generally, haploid plants are smaller in size (Dunwell, 2010; Smith et al., 1981; Norris et al., 1982) and have fewer guard cells. The ploidy status can also be determined by flow cytometry (Seguì-Simarro & Nuez, 2007; Biswas et al., 2011), using a Partec Cell Counter Analyzer (Ferrie et al., 2011) and chromosome counting (Biswas et al., 2011). Biswas et al., (2011) determined homozygosity in Valencia sweet orange doubled haploids using 43 simple sequence repeat (SSR) markers. Generally, an immersion of the small plant in 0.5% (w/v) colchicine for 24–48 h followed by a rinse in sterile, distilled water will double the chromosomes. Dhooghe et al. (2011) reviewed other antimitotic agents that have been used to double chromosomes.

A complete literature review of a specific plant species will usually provide a starting point to identify:

- the culture medium
- any specific growing conditions of the plant from which anthers will be harvested

- specific requirements for the stage of development of the microspore for optimal response
- any preculture treatments to enhance response
- specific light/dark (light spectrum) and temperature culture conditions
- necessary media changes during culture
- expected response rate
- separation of haploids, homozygous diploids, etc., and
- subsequent testing and identification of haploids and homozygous diploids.

A review of haploid technology can be found in Germanà (2011a,b).

The above very general comments provide an entry level base of knowledge to conduct the following exercises using anthers derived from plants that will respond in cell culture. These exercises using plant material that is widely available will establish whether or not an individual's skills in media preparation, explant surface disinfestation, and explant excision (careful excision of the anther to avoid bruising) are sufficient to successfully culture more challenging plant species.

DATURA ANTHER CULTURE

Purpose: To examine the effect of activated charcoal on the anther response of *Datura stramonium* Wright (common in central Texas). Most other *Datura* sp. will also give excellent results.

Medium Preparation: 1 liter equivalent, *Datura* Anther Medium.

1. Into a 1000-ml Erlenmeyer flask pour 500 ml of deionized, distilled water.
2. Mix in the following:
   a. 10 ml each Murashige and Skoog salts: nitrates, halides, NaFeEDTA, sulfates, and PBMo
   b. 10 ml myo-inositol stock (10 g/liter)
   c. 40 g sucrose
3. Adjust volume to 1000 ml. Adjust pH to 5.7.
4. Divide into two parts, 500 ml each:
   a. Add 4 g TC agar but not charcoal.
   b. Add 4 g TC agar and 1.5 g acid-washed charcoal [0.3%, (w/v)].
5. Cover flasks with aluminum foil.
6. Autoclave for 15 min at 121°C, 15 psi.
7. Distribute 25 ml per sterile plastic Petri dish (100 × 20 mm) by using a transfer hood.

Note: Variations in this experiment can include (a) using no agar and floating the anthers on the liquid medium and (b) cold treatments (4–12°C for 12–24 h).

Explant Preparation

Datura plants should flower in 8–10 weeks from seed germination, but allow 8–12 weeks for plant growth and flowering. Bud size is an indicator of the stage of microspore development. *Datura* buds between 4.6 and 6.0 cm are generally in the uninucleate stage and are most responsive.

Take a sample bud and place the anther on a slide; macerate in 1–2 drops of acetocarmine stain.

Push the large pieces of anther wall to the side and gently warm the side by passing it over a flame. Apply a cover slip. Observe the slide under a light microscope to determine the stage of microspore development.

Wash the buds in warm, soapy water. Disinfect in 5% chlorine bleach (2 drops Tween-20) for 10 min. Rinse three times in sterile water. Remove the calyx; cut the petal and peel back, exposing the anthers. Five anthers are in a bud. Gently sever the filament close to the base of the anther. Place cultures in the culture room. A response should be apparent in 4–8 weeks.

Questions

1. Discuss other factors that may influence anther response. How can anther response be increased?
2. How can you be sure a plant obtained by anther culture is indeed derived from the pollen?
3. What might be the effect of adding plant growth regulators to the medium?
4. What is the function of activated charcoal in the medium?

AFRICAN VIOLET

Purpose: To culture anthers of *Saintpaulia ionantha* H. Wendl. (African violet) for production of haploid plants and to observe the effect of ploidy on plant size and phenotype.

Medium Preparation: 1 liter equivalent, African Violet Anther Medium.

1. Into a 2000-ml Erlenmeyer flask pour 500 ml of deionized, distilled water.
2. Mix in the following:
   a. 10 ml each of the following stocks:

Compound	Stock	Weight of compound added
(A) KH_2PO_4	30 g/liter	300 mg
H_3BO_3	160 mg/liter	1.6 mg
(B) $MgSO_4 \cdot 7H_2O$	3.5 g/liter	35 mg
$MnSO_4 \cdot H_2O$	440 mg/liter	4.4 mg
$ZnSO_4$	150 mg/liter	1.5 mg
(C) KNO_3	100 g/liter	1000 mg
NH_4NO_3	100 g/liter	1000 mg
(D) Thiamine	10 mg/liter	0.1 mg
Glycine	200 mg/liter	2.0 liter
Nicotinic acid	50 mg/liter	0.5 mg
Pyridoxine	10 mg/liter	0.1 mg
(E) KI	80 mg/liter	0.8 mg
(F) Murashige and Skoog NaFeEDTA stock		

    b. 346 mg $Ca(NO_3)_2 \cdot 4H_2O$
    c. 64 mg KCl
    d. 30 g sucrose
    e. 50 ml IAA stock (10 mg/100 ml)
    f. 5 ml kinetin stock (10 mg/100 ml)
3. Adjust volume to 1000 ml. Adjust pH to 6.0.
4. Add 8 g TC agar. Melt.
5. Distribute 25 ml per culture tube (25 × 150 mm). Cap.
6. Autoclave for 15 min at 121°C, 15 psi.

Explant Preparation

Healthy African violet plants can be obtained in any season from most nurseries. Choose nonvariegated plant types because variegated plants do not seem to respond in culture. Collect unopened flower buds about 2–6 mm in length. Record the size of the flower buds. Wrap the buds in 10-cm cheesecloth squares and disinfect in 5% chlorine bleach for 2 min. Rinse with water.

In a sterile Petri dish, dissect out the immature anthers, being extremely careful not to bruise or damage the anthers. After the filaments are cut off and discarded, the anthers are ready to place in culture. Incubate on the culture shelf.

Observations

Small plants should be apparent in 7–8 weeks. Plants should be produced by 3–4% of the cultured anthers, but they may or may not be haploid. An examination of their chromosome is necessary to determine this.

If good technique is followed, contamination should be low (e.g., 0–20%).

After 2–3 months in culture the plants from the anther can be carefully separated and placed in a sterile Petri dish containing moistened vermiculite. These will undergo further growth. An alternative at this point is to separate the anther-derived plants and broken-off leaves and to culture these onto the rooting medium (MS salts, 2% sucrose 100 mg/liter myo-inositol, 0.6% agar). These will root and develop within 2–4 weeks.

The plants that were transferred into the Petri dishes with vermiculite at 4–6 weeks can be potted in an African violet potting mix.

Transfer of the anther-derived plants into the rooting medium usually results in 100% rooting of leaves and shoots.

Root Tip Chromosome Squash Technique

This technique can be used to determine the ploidy of African violet plants from anther culture. Roots obtained from leaf cuttings (such as African violets, rooted in distilled water of dilute nutrient solution for 2–3 weeks) can be used.

1. Place root tips in 5 N HCl for 10 min at room temperature.
2. Rinse three times in distilled water.
3. Gently macerate each root tip in a drop of 0.05% toluidine blue (Fisher T-161, C.1. No. 52040) buffered at pH 4.0 (0.1 M citric acid, 0.2 M Na_2HPO_4).

The tissue strains rapidly and is easily overstained. Squashes can be made and observed under the microscope.

Questions

1. Do you think varying the pH from 5.5 to 6.0 would affect anther response?
2. What would be the value of size reduction in a horticultural crop plant?
3. Do you think activated charcoal added to the medium would affect anther response?

TOBACCO

Purpose: To culture anthers from *Nicotiana* sp. to observe haploid plant development.

Medium Preparation: 1 liter equivalent. Tobacco Anther Medium.

1. Into a 2000-ml Erlenmeyer flask pour 500 ml of deionized, distilled water.
2. Mix in the following:
   a. 10 ml each Murashige and Skoog salts: nitrates, halides, NaFeEDTA, sulfates, and PBMo.
   b. 2.5 ml thiamine stock (40 mg/liter)
   c. 10 ml myo-inositol stock (10 g/liter)
   d. 30 g sucrose
   e. 2 mg glycine
   f. 10 ml vitamin stock
3. Adjust volume to 1000 ml. Adjust pH to 5.7.
4. Add 8 g TC agar or Difco-Bacto agar. Melt.
5. Distribute 25 ml per culture tube (25 × 150 mm). Cap.
6. Autoclave for 15 min at 121°C, 15 psi.

For an experimental variation, test 40 g sucrose and 0.1 mg/liter IAA.

Explant Preparation

Collect unopened flower buds at the developmental stage in which the corolla and calyx are of equal length (0.8–3 cm). This stage of flower bud development should have uninucleate microspores. Wrap the buds in 10-cm cheesecloth squares and disinfect in 20% chlorine bleach for 10 min. Rinse with water. Carefully isolate and culture the anther. Incubate on the culture shelf.

Observations

Carlson (1970) used late-tetrad-stage anthers of *Nicotiana tabacum* (L.) var. "Wisconsin 38" and obtained a 12% response from plated anthers.

To determine the ploidy of the anther-derived tobacco plants, the number of stomates on the leaf underside is an excellent indicator. Count the number of stomates on the diploid donor plant in a given leaf area. The haploid plants will have half this number of stomates. To count the stomates, a thin epidermal peel is torn off the underside of the leaf and immediately floated on a drop of water on a microscope slide. Add a coverslip and count the number of stomates in a given field of vision.

Questions

1. Why are there no plant growth regulators in this medium?
2. What other culture conditions could one examine to enhance anther response?
3. How might one double a haploid tobacco plant derived from anther culture?
4. What was the average number of stomates on the parent plant and from the anther-derived plants?

ANTHER SQUASH TECHNIQUE

To determine the stage of microspore (pollen) development within the anther it is necessary to examine the microspores.

1. Collect anthers from flower buds at various stages of development.
2. Place in a 3:1 solution of alcohol and acetic acid for 1–2 h or overnight.
3. Place anther on a clean slide; add a drop of 0.8% acetoorcein stain in 40% acetic acid. (Use 1% orcein in 50% acetic acid or 1 g orcein in 50 ml glacial acetic acid. Boil in a flask with a funnel on it and immediately remove from heat. Cool; add 50 ml deionized, distilled water. Stock: To 100 ml of preceding solution add 25 ml water.)
4. Gently macerate tissue with scalpel.
5. Remove large debris with scalpel.
6. Gently heat slide, being careful not to scorch it. (Touch the slide to the back of your hand to test how hot it is. If it is too hot for your hand, it is too hot.) As stain evaporates, add more; continue for 5–10 min. This allows nuclear material to take up stain.
7. Let slide sit for 5–10 min.
8. Place cover slip over material.
9. Place slide on paper towel and fold towel over slide; gently and *evenly* mash. Do not break the cover slip.
10. Observe under microscope.

BIBLIOGRAPHY

Anagnostakis, S. L. (1974). Haploid plants from anthers of tobacco—Enhancement with charcoal. *Planta*, *155*, 281–283.

Bayless, M. W. (1980). Cytology, chromosomal variation in plant tissue culture. In I. K. Vasil (Ed.), *International review of cytology: Perspectives in plant cell and tissue culture*. New York: Academic Press. (*Suppl. 11, Part A*).

Biswas, C. H., Amar, M. K. L., & XiuXin, M. H. (2011). Doubled haploid callus lines of Valencia sweet orange recovered from anther culture. *Plant Cell, Tiss., & Organ Cult.*, *104*(3), 415–423.

Burk, L. G., Gwynn, G. R., & Chaplin, J. F. (1972). Diploidized haploids from aseptically cultured anthers of *Nicotiana tabacum*. *Journal of Heredity*, *63*, 355–360.

Carlson, P. S. (1970). Induction and isolation of auxotrophic mutants in somatic cell cultures of *Nicotiana tabacum*. *Science*, *168*, 487–489.

Chia-chun, C., Tsun-wen, O., Hsu, C., Shu-min, C., & Chien-kang, C. (1978). A set of potato media for wheat anther culture. In *Proceedings of the Symposium on Plant and Tissue Culture* (pp. 51–55). Beijing Science Press China (subsidiary of VanNostrand–Reinhold, New York).

Collins, G. B. (1977). Production and utilization of anther-derived haploids in crop plants. *Crop Science*, *17*, 583–586.

Croser, J. S., Lülsdorf, M. M., Davis, P. A., Clarke, H. J., Bayliss, K. L., Mallikarjuna, N., & Siddique, K. H. M. (2006). Toward doubled haploid production in the Fabaceae: progress, constraints, and opportunities. *Critical Rev. in Planat Sci.*, *25*, 139–157.

Dhooghe, E., Van Laere, K., Eeckhaut, T., Leus, L., & Van Huylenbroeck, J. (2011). Mitotic chromosome doubling of plant tissues in vitro. *Plant Cell, Tiss., & Organ Cult.*, *104*, 359–373.

Dunwell, J. M. (1976). A comparative study of environmental and developmental factors which influence embryo induction and growth in cultured anthers of *Nicotiana tabacum*. *Environmental & Experimental Botany*, *16*, 109–118.

Dunwell, J. M. (2010). Haploids in flowering plants; origins and exploitation. *Plant Biotechnology J.*, *8*, 377–424.

Ferrie, A. M. R., & Mollers, C. (2011). Haploids and doubled haploids in Brassica spp. For genetic and genomic research. *Plant Cell, Tissue & Organ Culture*, *104*(3), 375–386.

Ferrie, A. M. R., Bethune, T. D., & Mykytyshyn, M. (2011). Microspore embryogenesis in Apiaceae. *Plant Cell Tiss. & Organ Cult.*, *104*, 399–406.

Germanà, M. A. (2011a). Anther culture for haploid and doubled haploid production. *Plant Cell Tiss. & Organ Cult.*, *104*, 283–300.

Germanà, M. A. (2011b). Gametic embryogenesis and haploid technology as valuable support to plant breeding. *Plant Cell Reports*, *30*(5), 839–857.

Gosal, S. S., Wani, S. H., & Kang, M. (2010). Biotechnology and crop improvement. *J. of Crop Improvement*, *24*(2), 153–217.

Guha, S., & Maheshwari, S. C. (1964). *In vitro* production of embryos from anthers of *Datura*. *Nature*, *204*, 497.

Guha, S., & Maheshwari, S. C. (1966). Cell division and differentiation of embryos in the pollen grains of *Datura in vitro*. *Nature*, *212*, 97–98.

Heberle-Bors, E., & Reinert, J. (1981). Environmental control and evidence for predetermination of pollen embryogenesis in *Nicotiana tabacum* pollen. *Protoplasma*, *109*, 249–255.

Hughes, K. W., Bell, S. L., & Caponetti, J. D. (1975). Anther-derived haploids of the African violet. *Canadian Journal of Botany*, *53*, 1422–1441.

Imamura, J., & Harada, H. (1980). Effects of abscisic acid and water stress on the embryo and plantlet formation in anther culture of *Nicotiana tabacum* Samsun. *Zeitschrift fuer Pflanzen-physiologie*, *100*, 285–289.

Johansson, L., Andersson, B., & Eriksson, T. (1982). Improvement of anther culture technique: Activated charcoal bound in agar medium in combination with liquid medium and elevated CO_2 concentration. *Physiologia Plantarum, 54*, 24–30.

Kim, K.-M., & Baensiger, P. S. (2005). A simple wheat haploid and doubled haploid production system using anther culture. *In Vitro Cell. Dev. Biol.-Plant, 41*, 22–27.

Luitel, S. S., & Won Hee, B. P. K. (2011). Agro-morphological characterization of anther derived plants in sweet pepper (*Capsicum annuum* L. cv. Boogie). *Horticulture, Environment and Biotechnology, 52*(2), 196–203.

Maheshwari, S. C., Rashid, A., & Tyagi, A. K. (1982). Haploids from pollen grains—Retrospect and prospect. *American Journal of Botany, 69*, 865–879.

Marks, G. E. (1973). A rapid HCl toludine blue squash technique for plant chromosomes. *Stain Technology, 48*, 229–231.

Martineau, B., Hanson, M. R., & Ausubel, F. M. (1981). Effect of charcoal and hormones on anther culture of *Petunia* and *Nicotiana*. *Zeitschrift feur Pflanzenphysiologie, 102*, 109–116.

McComb, J. A., & McComb, A. J. (1977). The cytology of plantlets derived from cultured anthers of *Nicotiana sylvestria*. *New Phytology, 79*, 679–688.

Mii, M. (1980). Effect of pollen degeneration on the pollen embryogenesis in anther culture of *Nicotiana tabacum* L. *Zeitschrift fuer Pflanzenphysiologie, 99*, 349–355.

Norris, R. E., Smith, R. H., & Turner, P. (1982). Phenotypic differences in haploid African violets. *In Vitro, 18*, 443–446.

Raghaven, V. (1997). Molecular embryology of flowering plants. (pp. 1–69). New York: Lambridge University Press.

Rashid, A., & Reinert, J. (1980). Selection of embryogenic pollen from cold-treated buds of *Nicotiana tabacum* var. Badischer Burley and their development into embryos in cultures. *Protoplasma, 109*, 161–167.

Rashid, A., & Reinert, J. (1981). *In vitro* differentiation of embryogenic pollen, control, control by cold treatment and embryo formation in the initiation of pollen cultures of *Nicotiana tabacum* var. Bradischer Burley. *Protoplasma, 109*, 285–294.

Reinert, J., & Bajaj, Y. P. S. (1977). Anther culture: Haploid production and its significance. In J. Reinert, & Y. P. S. Bajaj (Eds.), *Applied and fundamental aspects of plant cell, tissue, and organ culture* (pp. 251–264). New York: Springer-Verlag.

Reinert, J., Bajaj, Y. P. S., & Heberle, E. (1975). Induction of haploid tobacco plants from isolated pollen. *Protoplasma, 84*, 191–196.

Sangwan-Norreel, B. S. (1977). Androgenic stimulating factors in the anther and isolated pollen grain cultures of *Datura innoxia* Mill. *Journal of Experimental Botany, 28*, 843–852.

Segui-Simmarro, J. M., & Nuez, F. (2007). Embryogenesis induction, callogenesis, and plant regeneration by in vitro culture of isolated microspores and whole anthers. *J. Experimental Botany, 58*(5), 1119–1132.

Smith, R. H., Kamp, M., & Davis, R. S. (1981). Reduction in African violet size through haploidy. *In Vitro, 17*, 358–387.

Sopory, S. K., & Maheshwari, S. C. (1976). Development of pollen embryoids in anther cultures of *Datura innoxia*. II. Effects of growth hormones. *Journal of Experimental Botany, 27*, 58–68.

Sunderland, N. (1974). Anther culture as a means of haploid induction. In K. J. Kasha (Ed.), *Haploids in higher plants: Advances and potential* (pp. 91–122). Guelph, Ontario, Canada: Univ. of Guelph Press.

Sundar, A. N., & Jawahar, M. (2010). Efficient plant regeneration via somatic embryogenesis from anthers of *Datura stramonium* L. *Internat. J. of Agri. Tech., 6*(4), 741–745.

Sunderland, N., & Wicks, F. M. (1971). Embryoid formation in pollen grains of *Nicotiana tabacum*. *Journal of Experimental Botany, 22*, 213–226.

Sunderland, N., & Roberts, M. (1979). Cold-pretreatment of excised flower buds in float culture of tobacco anthers. *Annals of Botany*, *43*, 405–414.

Tomes, D. T., & Collins, G. B. (1976). Factors affecting haploid plant production from *in vitro* anther cultures of *Nicotiana* species. *Crop Science*, *16*, 837–840.

Tyagi, A. K., Rashid, A., & Maheshwari, S. C. (1981). Promotive effect of polyvinylpolypyrrol- idone on pollen embryogenesis in *Datura innoxia*. *Physiologia Plantarum*, *53*, 405–406.

Weatherhead, M. A., Bordon, J., & Henshaw, G. G. (1978). Some effects of activated charcoal as an additive to plant tissue culture media. *Zeitschrift fuer Pflanzenphysiologie*, *89*, 141–147.

Weatherhead, M. A., Grout, B. W. W., & Short, K. C. (1982). Increased hapoid production in *Saintpaulia ionantha* by anther culture. *Scientia Horticulturae*, *17*, 137–144.

Wernicke, W., & Kohlenbach, H. W. (1976). Investigation on liquid culture medium as a means of anther culture in *Nicotiana*. *Zeitschrift fuer Pflanzenphysiologie*, *79*, 189–198.

Embryo Rescue

Studies utilizing immature zygotic embryos help researchers gain greater insight into embryo development and seed maturation. Hanning (1940) was first to report plant development from isolated embryo culture. Embryo culture of immature orchid embryos (Knudson, 1922) has become important in obtaining rare hybrids. Embryo rescue has proven itself as a valuable tool for plant breeders to obtain hybrids from crosses that would otherwise abort on the plant.

There are numerous reports in the literature establishing the effectiveness of this technique in plant improvement programs. Immature heart stage and early cotyledonary stage embryos obtained from crossing *Salix* and *Populus* species were cultured on a half-strength Murashige and Skoog (1962) medium with 3% sucrose into plants (Bagniewska-Zadworna *et al.*, 2011). Plant material was confirmed to be hybrid using scanning electron microscopy, flow cytometry, and random amplification of polymorphic DNA screening. Bang *et al.* (2011) developed a novel CMS (cytoplasmic male sterility) line of radish using embryo rescue. Interspecific hybrids of cotton have been obtained by embryo rescue (Bhuyar *et al.*, 2009). Genovesi *et al.* (2009) obtained interploid hybrids from St. Augustine grass by embryo culture. Unique interspecific crosses in azalea, where there are both prezygotic and postzygotic barriers among interspecific crosses, were rescued by embryo culture (Eeckhaut *et al.*, 2007). Embryo rescue techniques were used to circumvent lengthy seed to seed cycle and fungal seed-borne diseases in globe artichoke breeding (Cravero & Cointry, 2007).

Plant Tissue Culture. Third Edition. DOI: 10.1016/B978-0-12-415920-4.00010-4
113

Some experimental parameters that are generally examined include deter-mining the stage at which abortion of the embryo occurs on the plant, and rescu-ing the embryo prior to that event. The longer an embryo can develop on the plant, the easier it is to isolate and successfully culture the embryo. Generally, the greater the number of days after pollination and before embryo abortion is considered optimal for embryo rescue. Sometimes plant growth regulators like abscisic acid, and gibberellin, kinetin and indole acetic acid are required. There can be variation in the number of responsive embryos depending on the species crossed (Bhuyar et al., 2009); only 9% of isolated azalea embryos were grown into seedlings (Eeckhaut et al., 2007). Behzad (2010) examined the number of days after pollination (55–60 days was optimum), strength of basal medium (half-strength was best), and plant growth regulators (1 mg/l NAA, 1 mg/l kinetin was chosen) on the rescue of interspecific Lupinus crosses.

The embryo can be excised from the plant and cultured in vitro into a viable plant. Very young zygotic embryos are difficult to culture in vitro; therefore, embryos beyond the globular stage are more successfully cultured. The develop-ment of a culture medium is usually not a complex process; a basic Murashige and Skoog (1962) basal medium with a carbohydrate in many cases will support embryo development into a plant.

The following exercises use plant material that is widely available in super-markets. The first laboratory exercise examines the effect of abscisic acid (ABA) on the developing sweet corn embryo. Abscisic acid (ABA) is a plant hormone that naturally occurs in seeds and is involved in embryo dormancy.

SWEET CORN EMBRYO CULTURE

Purpose: To demonstrate the effect of ABA on embryo development in vitro and provide the student with experience in embryo culture.

Medium Preparation: 1 liter equivalent, Corn Embryo Culture Medium.

1. Into a 1000-Erlenmeyer flask pour 500 ml of deionized, distilled water.
2. Mix in the following:
    a. 10 ml each Murashige and Skoog salts: nitrates, halides, NaFeEDTA, sulfates, and PBMo
    b. 10 ml thiamine stock (40 mg/liter)
    c. 40 g sucrose
3. Adjust volume to 800 ml.
4. Divide into four 200-ml portions labeled "A," "B," "C," and "D."
5. Add ABA stock (10 mg/100 ml) in the following amounts:
    a. no ABA
    b. 0.25 ml ABA stock
    c. 2.50 ml ABA stock
    d. 25.00 ml ABA stock
6. Adjust the volume of each flask to 250 ml.

7. Adjust pH to 5.7.
8. Add 2.0 g TC agar to each flask.
9. Autoclave for 15 min at 121°C, 15 psi. Also autoclave glass Petri dishes (100 × 20 mm) containing water-moistened filter paper to use as a sterile, moist surface on which to dissect the embryos.
10. Label the sterile plastic Petri dishes (60 × 15 mm) in a transfer hood using a marking pen:
    a. control
    b. 0.1 mg/liter ABA
    c. 1.0 mg/liter ABA
    d. 10.0 mg/liter ABA
11. Distribute 10 ml per sterile plastic Petri dish.

Preparation and Culture of Corn Embryos

1. Choose a fresh, disease-free ear of corn.
2. Remove silks and husks.
3. Remove individual whole kernels. Be very careful not to damage the pericarp.
4. Place the kernels in a beaker; cover the beaker with cheesecloth and put it under running water for 15 min.
5. While the kernels are being rinsed, practice dissecting out embryos from nonsterile kernels.
6. After 15 min, remove the kernels from the water and surface-disinfect in 20% chlorine bleach for 15 min.
7. Under the transfer hood, rinse three times in sterile water.
8. Carefully excise the embryo. Hold the kernel with forceps and use a scalpel to remove the pericarp. The pale yellow embryo should be apparent in the milky white endosperm (see Fig. 10.1). Culture one to 10 kernels per Petri dish.
9. Place the embryo on the culture medium. Place on the culture shelf for growth. Observe after 24, 48, and 72 h.

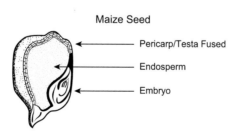

FIGURE 10.1   Hold the maize kernel with forceps and use a scalpel to remove the pericarp. The pale yellow embryo should be apparent in the milky white endosperm.

Questions

1. What effect did ABA have on the development of isolated corn embryos?
2. What function does ABA have in embryo development?

CRABAPPLE AND PEAR

Slow seed germination or seedling growth can delay breeding improvement programs. Nickell (1951) used embryo culture of weeping crabapple to shorten the time between seed ripening and seedling growth.

Purpose: To culture crabapple or ornamental pear embryos into seedlings and demonstrate the role of the seed coat in delaying germination.

Medium Preparation: 1 liter equivalent, Crabapple or Pear Embryo Culture Medium.

1. Into a 2000-ml Erlenmeyer flask pour 500 ml of deionized, distilled water.
2. Mix in the following:
   a. 10 ml each Murashige and Skoog salts: nitrates, halides, NaFeEDTA, sulfates, and PBMo
   b. 10 ml thiamine stock (40 mg/liter)
   c. 10 ml myo-inositol stock (10 g/liter)
   d. 20 g sucrose
3. Adjust volume to 1000 ml. Adjust pH to 5.7.
4. Add 8 g TC agar. Melt.
5. Distribute 25 ml per culture tube (25 × 150 mm). Cap.
6. Autoclave for 15 min at 121°C, 15 psi.

Explant Preparation

Collect ripened fruits from a tree in the fall. Wash with warm, soapy water. Wipe with 70% ethanol. Disinfect in 20% chlorine bleach for 15 min. Rinse three times in sterile water. In a sterile Petri dish cut the fruit in half to expose the seeds. Lift the seeds into a sterile Petri dish. Remove the seed coats to expose the embryo and culture. Culture some seeds with one or both of the cotyledons removed. Place the cultures on the culture shelf.

The seeds and embryos will take a lot of time to dissect out because the fruits are very hard. Practice outside the hood to gain experience in dissection of this material. Generally, very little to no contamination problems are encountered with these explants.

Observations

Growth of the embryos should be apparent in 48 h and after 2–3 weeks seedlings can be removed from the culture and established in soil.

Questions

1. Discuss the role of cotyledon in the development of the crabapple embryo.
2. Discuss why there was or was not an effect on germination when the cotyledons were removed.
3. Compare seedling growth (height) with and without cotyledons.

BIBLIOGRAPHY

Alvarez, M. N., Ascher, P. D., & Davis, D. W. (1981). Interspecific hybridization in *Euphaseolus* through embryo rescue. *HortScience, 16*, 541–543.

Arisumi, T. (1980). *In vitro* culture embryos and ovules of certain incompatible selfs and crosses among *Impatiens* species. *Journal of American Horticultural Science, 105*, 629–631.

Bagniewska-Zadworna, A., Zenkeler, M., Zenkeler, E., Wojciechowicz, M. K., Barakat, A., & Carlson, J. E. (2011). *Australian J. Bot., 59*(4), 382–392.

Bajaj, Y. P. S., Verma, M. M., & Dhanju, M. S. (1980). Barley × rye hybrids (Hordecale) through embryo culture. *Current Science, 49*, 362–363.

Bang, S. W., Tsutsui, K., Shim, S. H., & Kaneko, Y. (2011). Production and characterization of the novel CMS line of radish (*Raphanus sativus*) carrying *Brassica mauronum* cytoplasm. *Plant Breeding, 130*(3), 410–412.

Bennici, A., & Cionini, P. G. (1976). Cytokinins and *in vitro* development of *Phaseolus coccineus* embryos. *Planta, 147*, 27–29.

Behzad, H. (2010). Effects of the strength of basal medium and the plant growth regulators on the germination of ovules in interspecific crosses between L. x fomolongi x L. brownie. *World Applied Sci. J., 11*(4), 429–433.

Bhuyar, S. A., Sakhare, S. B., Patil, B. R., & Rathed, T. H. (2009). Embryo rescue studies in interspecific hybrids of cotton. *Annals of Plant Physiol., 23*(2), 272–273.

Bridgen, M. P. (1994). A review of plant embryo culture. *HortScience, 29*(11), 1243–1246.

Cravero, V., & Cointry, E. (2007). Shortening the seed-to-seed cycle in artichoke breeding by embryo culture. *Plant Breeding, 126*(2), 222–224.

Dolezel, J., Novak, F. J., & Luzny, J. (1980). Embryo development and *in vitro* culture of *Allium cepa* and its interspecific hybrids. *Zeitschrift fuer Pflanzenzuechtung, 85*, 177–184.

Eeckhaut, T. K., deHuylenbroeck, E., vanRiek, J., deBockstaele, J., & Van, E. (2007). Application of embryo rescue after interspecific crosses in the genus *Rhododendron*. *Plant Cell, Tiss. & Organ Cult, 89*(1), 29–35.

Genovesi, A. D., Jessup, R. W., Engelke, M. C., & Burson, B. L. (2009). Interploid St. Augustine grass (*Stenotaphrum secundatum* (Walt.) Kuntze) hybrids recovered by embryo rescue. *In Vitro Cell. & Dev. Bio.-Plant, 45*(6), 659–666.

Gill, B. S., Waines, J. G., & Sharma, H. C. (1981). Endosperm abortion and the production of viable *Aegilops squarrosa* × *Triticum boecticum* hybrids by embryo culture. *Plant Science Letters, 23*, 181–187.

Goldstein, C. S., & Kronstad, W. E. (1986). Tissue culture and plant regeneration from immature embryo explants of barley. *Hordeum vulgare. Theoretical and Applied Genetics, 71*, 631–636.

Hannig, E. (1904). Physiologie pflanzliche Embryonen. I. Ueber die Kultur on Cruciferen-Embryonen ausserhalb des Embryosacks. *Botanishe Zeitung, 62*, 45–80.

Knudson, L. (1922). Nonsymbiotic germination of orchid seeds. *Botanical Gazette, 73*, 1–25.

Ladizinsky, G., Cohen, D., & Muehlbauer, F. J. (1985). Hybridization in the genus *Leas* by means of embryo culture. *Theoretical and Applied Genetics, 70*, 97–101.

Laibach, F. (1929). Ectogenesis in plants: Methods and genetic possibilities of propagating embryos otherwise dying in the seed. *Journal of Heredity, 20*, 200–208.

Mauney, J. R. (1961). The culture *in vitro* of immature cotton embryos. *Botanical Gazette, 122*, 205–209.

Murashige, T., & Skoog, F. (1962). A revised medium for rapid growth and bioassays with tobacco tissue cultures. *Physiologia Plantarum, 15*, 473–497.

Narayanaswami, S., & Norstog, K. (1964). Plant embryo culture. *Botany Review, 30*, 587–628.

Nickell, L. G. (1951). Embryo culture of weeping crabapple. *Proceedings of the American Society of Horticultural Science, 57*, 401–405.

Norstog, K., & Klein, R. M. (1972). Development of cultured barley embryos. II. Precocious germination and dormancy. *Canadian Journal of Botany, 50*, 1887–1894.

Phillips, G. C., Collins, G. B., & Taylor, N. L. (1982). Interspecific hybridization of red clover (*Trifolium pratense* L.) with *T. sarosiense* Hazsl. Using *in vitro* embryo rescue. *Theoretical and Applied Genetics, 62*, 17–24.

Raghavan, V. (1977). Applied aspects of embryo culture. In J. Reinert, & Y. P. S. Bajaj (Eds.), *Applied and fundamental aspects of plant cell, tissue, and organ culture* (pp. 375–397). New York: Springer-Verlag.

Raghavan, V. (1980). Embryo culture. In I. K. Vasil (Ed.), *International review of cytology: Advances in plant cell and tissue culture*. New York: Academic Press. (*Suppl. 11, Part B*, pp. 209–240).

Raghavan, V., & Torrey, J. G. (1964). Effects of certain growth substances on the growth and morphogenesis of immature embryos of Capsella in culture. *Plant Physiology, 39*, 691–699.

Rangan, T. S., Murashige, T., & Bitters, W. P. (1968). *In vitro* initiation of nucellar embryos in monoembryonic *Citrus*. *HortScience, 3*, 226–227.

Rappaport, J. (1954). *In vitro* culture of plant embryos and factors controlling their growth. *Botany Review, 20*, 201–225.

Rietsema, J., Satina, S., & Blakeslee, A. F. (1953). The effect of sucrose on the growth of Datura stramonium embryos *in vitro*. *American Journal of Botany, 40*, 538–545.

Sinska, I., & Lewak, S. (1977). Is the gibberellin A_4 biosynthesis involved in the removal of dormancy in apple seeds? *Plant Science Letters, 9*, 163–170.

Stafford, A., & Davies, D. R. (1979). The culture of immature pea embryos. *Annals of Botany, 44*, 315–321.

van Overbeek, J., Conklin, M. E., & Blakeslee, A. F. (1942). Cultivation *in vitro* of small *Datura* embryos. *American Journal of Botany, 29*, 472–477.

Williams, E. G., & de Lautour, G. (1980). The use of embryo culture with transplanted nurse endosperm for the production of interspecific hybrids in pasture legumes. *Botanical Gazette, 14*, 252–257.

Meristem Culture for Virus-Free Plants

Many plants are internally infected with viruses, and this results in less vigorous growth, necrosis, curling of leaves, streaks in leaves or flowers, decreases in yield, and plant death (Quak, 1977). There is no chemical treatment to rid a plant of virus infection; however, viruses are not generally spread to the progeny through seed. In clonally propagated plants like potato, garlic, pineapple, orchids, carnation, banana, and strawberry, virus contamination can be high. Two methods to free vegetatively propagated plants of virus are thermotherapy and meristem-tip culture. The shoot apical meristem and first set of primordial leaves in an elongating shoot are generally not connected to the vascular system of the plant and, therefore, are not contaminated by viruses that travel through the vascular system. Morel and Martin (1952) did the pioneering work on the establishment of virus-free dahlias using apical meristem culture. If this explant is carefully excised so as not to contaminate it with sap from more mature leaves or stem tissue and is placed in culture, a virus-free plant can be established. Many important horticultural and agronomically important crops are routinely freed of viral contamination using this procedure. The literature abounds with

Plant Tissue Culture. Third Edition. DOI: 10.1016/B978-0-12-415920-4.00011-6

reports of specific plant species and successful establishment of plants free of virus infestation.

Thermotherapy or heat treatment is sometimes used alone or in combination with meristem culture to rid potato cultivars of viruses (Zapata *et al.,* 1995; Zapata and Smith, 1998). Plants can be placed in a growth chamber or placed in hot water to administer the heat treatment. Heat will inactivate the virus (Matthews, 1991). The combination of heat and meristem culture of lilies is also described by Nesi *et al.* (2011) to free bulbs of two viruses.

Meristem culture has been successfully used to free plants of viroids (smaller than viruses) and plant pathogens. Banerjee *et al.* (2010) reported freeing an *Artemisia* species of a phytoplasm utilizing apical meristem culture. The resulting plants were tested using PCR, visual microscopic examination, and visual avaluation of morphological features to be free of the phytoplasma. Strawberry plants were freed of the crown rot pathogen through meristem culture (Whitehouse *et al.*, 2011). Meristem culture was reported to free plants of viroid infection (Hosokawa, 2008).

The process of meristem culture does not always free the plants from viruses; therefore, claims of virus-free plants must be confirmed. In order to verify the virus status of a plant, analytical techniques such as ELISA (a serological technique to identify viruses: enzyme-linked immunosorbent assay), DAS-ELISA (double antibody sandwich-enzyme-linked immunosorbent assay), and RT-PCR (reverse transcription-polymerase chain reaction: a molecular technique) are used (Myouda *et al.*, 2005; Shiraqi *et al.*, 2010; Nesi *et al.*, 2011; Youssef *et al.*, 2009).

There is confusion in the literature regarding the term "meristem culture." The true shoot meristem consists only of the isolated apical dome without visible primordial leaves attached. This is a very difficult explant to culture because of its small size. Smith and Murashige (1970) reported the first true meristem culture of an isolated angiosperm meristem into a complete plant. Before that time it was believed that the isolated shoot apical meristem of an angiosperm could not direct its own development but, rather, relied on subjacent primordial leaves and stem tissue (Ball, 1946, 1960).

Generally, to establish a virus-free plant one can culture the apical dome plus two to four subjacent primordial leaves. For enhanced probability of survival, choose a shoot tip that is in a stage of rapid growth; such a shoot tip is usually not virus infested. The virus is usually in the vascular system and in a rapidly growing shoot, the vascular system is not yet connected to the small shoot tip. The isolation of the meristem and primordial leaves is a very tedious operation requiring patience and knowledge of the anatomy and location of this microscopic explant. Practice in isolation of this tissue is required before successful cultures can be established. Care must be taken during isolation not to damage the meristem or allow it to dry. Final cuts have to be made by very fine instruments.

Razor blade slivers mounted in a chuck handle have worked for *Coleus* meristem excision (Smith and Murashige, 1970).

ISOLATION OF THE SHOOT APICAL MERISTEM

Purpose: To gain experience in the isolation of a shoot apical meristem.

   Medium Preparation: 1 liter equivalent, *Coleus* (Smith and Murashige, 1982) and *Tropaeolum* (Ball, 1946) Meristem Medium.

1. Into a 2000-ml Erlenmeyer flask pour 500 ml of deionized distilled water.
2. Mix in the following:
    a. 10 ml each Murashige and Skoog salts: nitrates, halides, NaFeEDTA, sulfates, and PBMo
    b. 10 ml thiamine stock (40 mg/liter)
    c. 10 ml myo-inositol stock (10 g/liter)
    d. 30 g sucrose
    e. 10 ml IAA stock (10 mg/100 ml)
3. Adjust volume to 1000 ml. Adjust pH to 5.7.
4. Add 8 g TC agar.
5. Autoclave for 15 min at 121°C.
6. Distribute 25 ml per sterile plastic Petri dish (100 × 20 mm).

Explant Preparation

Collect shoot tips of *Tropaeolum majus* L. or *Coleus blumei*. Wash the tips in warm, soapy water. Remove large leaves. Wrap the tips in cheesecloth; disinfect in 10% chlorine bleach for 10 min. Rinse three times in sterile water. To avoid dessication of delicate tissue during isolation, place tissue on a sterile water-moistened filter paper in a Petri dish. Under a microscope, isolate the shoot apical dome. Do not include visible leaf primordial (see Fig. 11.1).

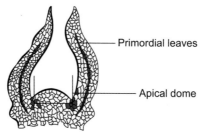

— Primordial leaves

— Apical dome

FIGURE 11.1   The explant for meristem culture is the shoot apical dome. Do not include visible leaf primordia.

Questions

1. What might explain why this explant does not generally harbor virus infection?
2. Do you think there would be a difference in the success rate of this technique if dormant instead of flush growth shoot tips were used?
3. What is the main reason for lack of growth of this explant?
4. Does this technique guarantee a virus-free plant?
5. What is virus indexing, and what does it tell you?
6. Observe the apices under a dissecting microscope. Do they disorganize to form callus? What enlarges first?

ESTABLISHING VIRUS- AND BACTERIA-FREE PLANTS

Purpose: To culture very small shoot apices or apical meristems of *Diffenbachia* (Knauss, 1976) to obtain bacteria- and virus-free plants.

Medium Preparation: 1 liter equivalent, *Diffenbachia* Meristem Medium.

1. Into a 2000-ml Erlenmeyer flask pour 500 ml of deionized, distilled water.
2. Mix in the following:
   a. 10 ml each Murashige and Skoog salts: nitrates, halides, NaFeEDTA, sulfates, and PBMo
   b. 10 ml thiamine stock (40 mg/liter)
   c. 10 ml myo-inositol stock (10 g/liter)
   d. 30 g sucrose
   e. 10 ml $NaH_2PO_4 \cdot H_2O$ stock (17 g/liter)
   f. 160 ml 2iP stock (50 mg/500 ml)
   g. 20 ml IAA stock (10 mg/100 ml) h. 10 ml vitamin stock
   i. 2 mg glycine
   j. 8 ml adenine sulfate stock (1 g/100 ml)
3. Adjust volume to 1000 ml. Adjust pH to 5.7.
4. Add 8 g TC agar.
5. Autoclave for 15 min at 121°C, 15 psi.
6. Distribute 25 ml per sterile plastic Petri dish (100 × 20 mm).

Explant Preparation

Remove the outer leaves of the shoot tips. Wash the shoot tips in warm, soapy water. Place the tips in a beaker covered by cheesecloth and place the beaker under running tap water for 2–3 h. Wrap the explant in cheesecloth, disinfect in 10% chlorine bleach for 15 min, and rinse three times in sterile water. As you peel down the explant, dip your instruments in alcohol, flame them, and then *cool* them. Dip the explants in 5% chlorine bleach followed by sterile water before planting. Place the cultures on the culture shelf.

GARLIC PROPAGATION FROM BULB SCALES

Garlic is a valuable commercial crop that can only be produced vegetatively because it is sterile. Many commercial garlic varieties are virus infected, and result in reduced yields. Therefore, tissue culture is a useful method for producing virus-free garlic.

Purpose: Clonal propagation and establishment of virus-free garlic (*Allium sativum* L.) (AboEl-Nil, 1977; Bhojwani, 1980; Novàk, 1983) from clove explants.

Medium Preparation: 1 liter equivalent, Garlic Bulb Scale Medium.

1. Into a 2000-ml Erlenmeyer flask pour 500 ml of deionized, distilled water.
2. Mix in the following:
   a. 10 ml each Murashige and Skoog salts: nitrates, halides, NaFeEDTA, sulfates, and PBMo
   b. 2.5 ml thiamine stock (40 mg/liter)
   c. 100 my myo-inositol stock (10 g/liter)
   d. 30 g sucrose
   e. 10 ml vitamin stock
   f. 5 ml 2iP stock (10 mg/100 ml)
   g. 2.4 ml NAA stock (10 mg/100 ml)
3. Adjust volume to 1000 ml. Adjust pH to 5.7.
4. Add 8 g TC agar. Melt.
5. Distribute 25 ml per culture tube (25 × 150 mm). Cap.
6. Autoclave for 15 min at 121°C, 15 psi.

Explant Preparation

1. Using garlic from the supermarket (12–20 cloves per bulb), separate the cloves from the bulbs.
2. Disinfect and section clove (see Fig. 11.2)
   a. Soak each clove in 95% ethanol solution for 30 s.
   b. Flame.
   c. Place in a sterile dish.
   d. Remove scorched leaf envelope and scorched root plate.
   e. Split open fleshly storage leaf.
   f. Excise shoot from disk; remove the outer leaf and plant shoot in medium.
3. Culture on culture shelf.

Data

1. How many plants were obtained for an explant? What was the average number of plants obtained?
2. From what area did the shoots arise?

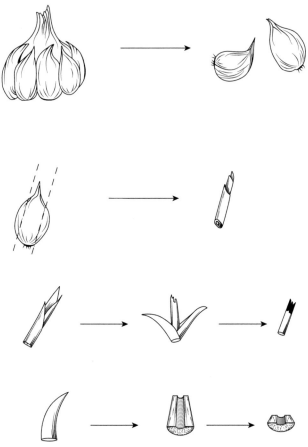

FIGURE 11.2   Follow these steps for the clonal propagation and establishment of virus-free garlic.

BIBLIOGRAPHY

AboEl-Nil, M. M. (1977). Organogenesis and embryogenesis in callus cultures of garlic (*Allium sativum* L.). *Pl. Sci. Lettr.*, *9*, 259–264.

Ball, E. (1946). Development in sterile culture of stem tips and subjacent regions of *Tropaeolum majus* L. and *Lupinus albus* L. *Amer. J. Bot.*, *33*, 301–318.

Ball, E. (1960). Sterile culture of the shoot apex of *Lupinus albus* L. *Growth*, *24*, 91–110.

Banerjee, S., Haider, R., Bagchi, G. D., & Samad, A. (2010). Regeneration of phytoplasma-free *Artemisia roxburghiana* Besser var. *purpurascens* (Jacq.) Hook. Plants using apical meristem culture. *Plant Cell, Tiss. & Organ Cult.*, *103*(2), 189–196.

Bhojwani, S. S. (1980). *In vitro* propagation of garlic by shoot proliferation. *Scientia Horticulturae*, *13*, 47–52.

Bhojwani, S. S., & Razdan, M. K. (1986). *Plant tissue culture: Theory and practice*. New York: Elsevier.

Brettell, R. I. S., & Ingram, D. S. (1979). Tissue culture in the production of novel disease-resistant crop plants. *Biological Rev.*, *54*, 329–345.

Gamborg, O. L. (2002). Plant tissue culture. Biotechnology milestones. *In Vitro Cell. Dev. Biol.-Plant*, *38*, 84–92.

George, F. F., & Sherrington, P. D. (1984). *Plant propagation by tissue culture*. Eversley, England: Exegtics.

Hollings, M. (1965). Disease control through virus-free stock. *Ann. Rev. Phytopath.*, *3*, 367–396.

Hosokawa, M. (2008). Leaf primordial-free shoot apical meristem culture: A new method for production of viroid-free plants. *Japanese Soc. Hort. Sci.*, *77*(4), 341–349.

Kehr, A. E., & Schaeffer, G. W. (1976). Tissue culture and differentiation of garlic. *HortScience*, *11*, 422–423.

Knauss, J. F. (1976). A tissue culture method for producing *Diffenbachia picta* cv. "Perfection" free of fungi and bacteria. *Proc. Florida State Hort. Soc.*, *89*, 293–296.

Matthews, R. E. F. (1991). Virus-free seed. In R. E. F. Matthews (Ed.), *Plant virology* (pp. 599–630). New York: Academic Press.

Morel, G. (1960). Producing virus-free Cymbidiums. *American Orchid Society Bulletin*, *29*, 495–497.

Morel, G. (1975). Meristem culture techniques for the long-term storage of cultivated plants. In O. H. Frankel, & J. G. Hawkes (Eds.), *Crop genetic resources for today and tomorrow* (pp. 327–332). New York: Cambridge Univ. Press.

Morel, G., & Martin, C. (1952). Guerison de dahlias atteints d'une maladie à virus. *Comptes Rendus Hebdomadaires des Seances de l'Academie des Sciences (Paris)*, *234*, 1324–1325.

Myouda, T., Sanada, A., Fuji, S., Natsuaki, K. T., Koshio, K., Toyohara, H., Kikuchi, F., & Fujimaki, H. (2005). Propagation of greater yam (*Dioscorea alata* L.) using shoot apex and nodal segment cultures combined with virus detection by RT-PCR. *Sabrao J. of Breeding & Gen.*, *37*(1), 65–70.

Nehra, N. S., & Kartha, K. K. (1994). Meristem and shoot tip culture: Requirements and applications. In I. K.Vasil, & T. A. Thorpe (Eds.), *Plant cell and tissue culture* (pp. 37–71). Amsterdam: Kluwer Academic Publishers.

Nesi, B., Lazzereschi, S., Pecchiolos, S., & Grasotti, A. (2011). Virus detection and propogation of virus-free bulbs from selected progenies of pollenless lilies. *Acta Horticulturae*, *900*, 325–331.

Nishi, S., & Ohsaua, K. (1973). Mass production method of virus-free strawberry plants through meristem callus. *Japanese Agri. Research Quarterly (Tokyo)*, *7*, 189–194.

Novàk, F. J. (1972). The changes in karotype in callus culture of *Allium sativum* L. *Caryologia*, *27*, 45–51.

Novàk, F. J. (1983). Production of garlic (*Allium sativum* L.) tetraploids in shoot-tip *in vitro* culture. *Zeitschrift fuer Pflanzenzuechtung*, *91*, 329–333.

Quak, F. (1977). Meristem culture and virus-free plants. In J. Reinert, & Y. P. S. Bajaj (Eds.), *Applied and fundamental aspects of plant cell, tissue, and organ culture* (pp. 598–646). New York: Springer-Verlag.

Raychaudhuri, S. P., & Verma, J. P. (1977). Therapy by heat, radiation and meristem culture. In J. G. Horsfall, & E. Bowling (Eds.), *Plant disease: An advanced treatise* (pp. 177–189). New York: Academic Press.

Shabde, M., & Murashige, T. (1977). Hormonal requirements of excised *Dianthus caryophyllus* L. shoot apical meristem *in vitro*. *Amer. J. Bot.*, *64*, 443–448.

Shiraqi, M. H. K., Baque, M. A., & Nasiruddin, K. M. (2010). Eradication of banana bunchy top virus (BBTV) through meristem culture of infected plant banana cv. *Sabri. Hort., Envir., & Biotech.*, *51*(3), 212. 221.

Smith, R. H., & Murashige, T. (1970). *In vitro* development of isolated shoot apical meristems of angiosperms. *Amer. J. Bot.*, *57*, 562–568.

Smith, R. H., & Murashige, T. (1982). Primordial leaf and phytohormone effects on excised shoot apical meristems of *Cloeus blumei* Benth. *Amer. J.f Bot.*, *69*, 1334–1339.

Smith, R. H., & Park, S. H. (2001). Tissue culture for crop improvement. In M. S. Kang, (Ed.), *Quantitative Genetics, Genomics, and Plant Breeding* (pp. 189–196). New York: CABI.

Thorpe, T. A. (2007). History of plant tissue culture. *Mol. Biotechnol.*, *37*, 169–180.

Whitehouse, A. B., Govan, C. L., Hammond, K. J., Sargent, D. J., & Simpson, D. W. (2011). Meristem culture for the elimination of the strawberry crow rot pathogen, *Phytophthora cactorum*. *J. Berry Research*, *1*(3), 129–136.

Youssef, S. A., Al-Dhaher, M. M. A., & Shalaby, A. A. (2009). Elimination of grapevine fanleaf virus (GFLV) and grapevine leaf roll-associated virus-1 (GLRaV-1) from infected grapevine plants using meristem tip culture. *Internat. J. Virology*, *5*(2), 89–99.

Zapata, C., Miller, C., & Smith, R. H. (1995). An in vitro procedure to eradicate potato viruses X, Y, and S from Russet Norkotah and two if its strains. *In Vitro Cell. & Dev. Bio.-Plant*, *31*, 153–159.

Zapata, C., & Smith, R. H. (1998). *Freeing plants of viruses*. In J. E. Celis, (Ed.), *Cell biology: A laboratory handbook* (pp. 455–489). New York: Academic Press.

Chapter 12

In Vitro Propagation for Commercial Production of Ornamentals

The widespread interest in using tissue culture to mass clone ornamental plants began in the 1960s. A gerbera daisy grower in southern California visited Dr. Murashige at the University of California, Riverside campus. The grower's problem in gerbera daisy production was that through seed propagation, a mixture

Plant Tissue Culture. Third Edition. DOI: 10.1016/B978-0-12-415920-4.00012-8

of flower colors resulted; this made production of specific flower colors difficult. Working with the grower, Murashige developed a protocol for gerbera clonal propagation. Subsequently, Dr. Murashige started workshops to help train other growers in development of propagation protocols for specific ornamentals. In return, growers funded symposium and micropropagation sessions at the annual meetings of the Society for In Vitro Biology. Soon ornamental producers all over the United States were setting up tissue culture labs.

Murashige (1974) identified major stages in the *in vitro* propagation process. Stage I is the establishment of aseptic cultures. This stage can often be difficult due to contamination and production of phenolic compounds by the explant. However, in this first stage, the goal is to optimize a surface disinfestation protocol and nutrient medium for survival and growth of the explant.

Stage II culture is for multiplication of a large number of shoots. Generally, a cytokinin enhances multiple-shoot production from preexisting axillary buds or multiplication is achieved by adventitious bud formation from leaf, stem, or petiole explants. Somatic embryo formation can also result in a high multiplication rate. Stage II cultures can be subcultured three to six times. More subcultures increase the chance of undesirable off-type plants. Reculture of Stage II material can average 5–10 plants or more per explant every 6–8 weeks. If callus is involved in this stage, the occurrence of aberrant plants or somaclonal variation is a problem. Axillary shoot multiplication is preferred for Stage II multiplication because the risk of producing aberrant plants is extremely low as compared to going through a callus stage.

Stage III prepares plants for transfer to soil. Generally an auxin is included for rooting of individual shoots. For many plants, this stage can take place in the greenhouse in potting soil. A publication by Oinam *et al.* (2011) gives a general overview of adventitious root initiation. Also refer to De Klerk (2002) for rooting of microcuttings.

Stage IV is now a recognized step in propagation. In this stage, the rooted cuttings are removed from the culture tube, and the agar is gently washed from the root system. The plants are placed in a potting mix and kept in high humidity and shade. Usually after 2 weeks, the plants have been conditioned and can tolerate more light and lower humidity.

In vitro propagation, or micropropagation, has advantages over conventional propagation for many, but not all, plants. The major benefit is clonal propagation, or asexual reproduction, resulting in genetically identical copies of a cultivar. Desirable stock plants can be chosen, and thousands of identical clones can be produced. Other benefits of *in vitro* propagation include more rapid propagation, maintenance of pathogen-free stock plants, enhanced axillary branching of *in vitro*-derived plants (resulting in fuller foliage), earlier flowering, production of a uniform crop, year-round production, hastening of a new crop introduction, and cloning of only the desirable female plants (date palm) or male plants (asparagus).

Vitrification or hyperhydricity can be a problem in micropropagation (Rojas-Martinez *et al.*, 2010; Ivanova & Van Staden, 2010; Kevers *et al.*, 2004). Shoots that are in this condition are difficult to establish as plants. *In vitro* shoots have a glass-like appearance. Hyperhydricity is characterized by large intercellular spaces, less epicuticular wax, fewer stomata on leaves, chloroplasts with small granna and a lack of starch grains. Preece (2010) reviews this condition discussing gelling agents and liquid media culture systems to enhance micropropagation of many plant species and effects on hyperhydricity.

Literature on micropropagation is replete with examples of improved cell culture protocols for a specific species or cultivar (see the Bibliography at the end of the chapter). At the time of writing (in 2012), there is still not a single, simple protocol that can be universally applied to *in vitro* propagate different plant species or sometimes even different cultivars of a single species. The literature is a starting point, and having prior experience in the stages of propagation and cell culture is important. The exercises that follow demonstrate some of the various techniques of *in vitro* propagation. Many general references on *in vitro* propagation of different plant species are included.

BOSTON FERN

Many species of ferns are commercially important for their uses in landscaping, as houseplants, and in floral arrangements. Classically, ferns have been propagated either from spores or by division. Although propagation from spores can result in the formation of many plants, it can be a slow process.

Ferns have two phases, referred to as alternation of generations, in their vegetative and reproductive growth cycle. Spore germination results in a layer of small, moss-like clumps that contain male and female organs. Fertilization of these gametophytes results in the formation of sporophytes or spore-producing plants. It generally takes 2–4 months from spore germination for sporophytic plants to develop.

Division can be used to propagate species that develop rhizomes and form offsets. This is an easy procedure; however, the number of propagules is limited, and many species do not have this type of growth.

The Boston fern (*Nephrolepsis exaltata* Schott. cv. *bostoniensis*) belongs to the family Polypodiaceae, a genus that includes approximately 30 species from tropical areas of both the New and Old Worlds. The Boston fern originated from a Florida wild sword fern, *Nephrolepis exaltata*. In 1894 a shipment of sword ferns sent from a Philadelphia grower to a Boston distributor was different from the true *N. exaltata*. In 1896 the new fern was named *N. exaltata* cv. *bostoniensis*.

The Boston fern is widely cultivated and is a popular houseplant. Generally, the ferns are propagated in large planting beds. The thin, trailing shoots (called runners) root, forming numerous plants on the surface of the bed. Unfortunately, plants propagated in this way show a lack of uniformity and often fall prey to parasites.

In vitro propagation of Boston ferns has many advantages (Burr, 1976; Harper & Murashige, 1976; Murashige, 1974; Oki, 1981; Roberts, 1965; Wetherell, 1982). The resulting plants are identical to the selected mother plant and are thus uniform in both size and appearance. Ferns cultivated *in vitro* also tend to be bushier and more attractive than those propagated in planting beds. As a special bonus to the commercial grower, tissue culture produces a proliferation of ferns that cannot be equaled by conventional methods.

Stage I

Purpose: To initiate aseptic cultures of Boston fern runner tips and to observe principles of rapid clonal propagation.

    Medium Preparation: 150 ml equivalent, Stage I Boston Fern Medium.

1. Into a 250-ml Erlenmeyer flask pour 60 ml of deionize, distilled water.
2. Mix in the following:
   a. 1.5 ml each Murashige and Skoog salts: nitrates, halides, NaFeEDTA, sulfates, and PBMo
   b. 1.5 ml thiamine stock (40 mg/liter)
   c. 1.5 ml myo-inositol stock (10 g/liter)
   d. 4.5 g sucrose
3. Adjust volume to 150 ml. Adjust pH to 5.7.
4. Dispense 5-ml aliquots to 30 culture tubes (25 × 150 mm).
5. Autoclave for 15 min at 121°C, 15 psi.

Explant Preparation

Cultures of Boston fern runner tips are less likely to become contaminated than are cultures of most other explants. This difference is largely due to the fact that the runner tips are usually not in contact with the soil (many are in hanging baskets) and are, therefore, relatively contaminant-free before culture initiation. To minimize the possibility of culture contamination, place your stock plant in a cool, dry environment for several weeks before explant removal. Air-conditioned rooms are perfect for this purpose.

    Always handle the runner tips with great care. Damaged plant material will not grow in the culture medium.

    Choose a healthy plant that has numerous runners growing from the crown of the fern fronds. Remove actively growing runner tips covered with whitish hairs about 3 cm long from the plant. Wash in warm, soapy water and rinse. In batches of about 12, wrap the runner tips in cheesecloth squares so that they do not float out of the container while you are surface-disinfecting and rinsing. Sterilize the tips in 10% chlorine bleach for 10 min and rinse three times in sterile water. With forceps and a scalpel, remove the terminal 1.5 cm and plant it, one to three runners per test tube (see Fig. 12.1). This is delicate tissue; do not pinch, bruise, or squash it with the forceps. Place the cultures on the culture shelf.

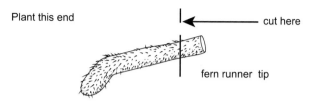

Plant this end

cut here

fern runner tip

FIGURE 12.1   Plant the terminal 1.5 cm of the Boston fern runner tip, taking care not to pinch, bruise, or crush it with the forceps.

Observations

After 1 week record and remove contaminated cultures. After 6–8 weeks record the number of plants produced per runner tip. Each of these single plants will be subcultured in Stage II for multiplication and further root growth.

Boston fern fronds proliferate best in a liquid Stage I medium. As an additional advantage, the omission of agar reduces the cost of medium preparation and speeds test tube clean-up.

After about 6–8 weeks you should be able to go to Stage II.

Stage II

The objective of Stage II is proliferation of shoots. In this stage it is important to select healthy Stage I cultures and to subculture for rapid multiplication. Shoots with poor or slow growth should be discarded because plants derived from these may not be strong. If shoot material from Stage II is subcultured in the Stage II medium for an increase in the number of plants, this should not be done more than four times because repeated subcultures may produce abnormal plants. Sodium phosphate (NaH_2PO_4 . H_2O) has been added to the Stage II medium to enhance shoot development.

Purpose: To subculture Stage I fern runner tips onto the Stage II medium and observe multiplication rates.

Medium Preparation: 1 liter equivalent, Stage II Boston Fern Medium.

1. Into a 2000-ml Erlenmeyer flask pour 500 ml of deionized, distilled water.
2. Mix in the following:
   a. 10 ml each Murashige and Skoog salts: nitrates, halides, NaFeEDTA, sulfates, and PBMo
   b. 10 ml thiamine stock (40 mg/liter)
   c. 10 ml myo-inositol stock (10 g/liter)
   d. 30 g sucrose
   e. 1.0 ml NAA stock (10 mg/100 ml)
   f. 10 ml kinetin stock (10 mg/100 ml)
   g. 10 ml NaH_2PO_4 . H_2O stock (17 g/liter)
3. Adjust volume to 1000 ml. Adjust pH to 5.7.
4. Add 8 g TC agar. Melt.

5. Distribute 25 ml per culture tube (25 × 150 mm). Cap.
6. Autoclave for 15 min at 121°C, 15 psi.

Explant Preparation

Use Stage I explants. Transfer them to a sterile Petri dish and divide.

Observations

In 3–5 weeks you should observe the growth of more shoots from each of the cultures. In 4–6 weeks the shoots should develop rhizomes (brown underground stems). During these weeks, the cultures will also undergo shoot proliferation and frond elongation.

If large numbers of plants are desired, Stage II can be repeated three or four times. After 7 weeks of growth divide each rhizome clump into two pieces and trim fronds to a length of 15–20 mm. It is very important to leave some green material on each clump. Place each clump into a tube of fresh Stage II medium. Each successive culture will yield additional plant material.

Stage III

The goal of Stage III is the preparation of plant material for transplantation to soil. When working with Boston ferns, it is possible to omit this preparatory stage and transfer the shoots directly from the Stage II medium to a potting mix. However, including this step ensures strong root development.

Purpose: To culture Stage III of the Boston fern.

Medium Preparation: 1 liter equivalent, Stage III Boston Fern Medium.

1. Into a 2000-ml Erlenmeyer flask pour 500 ml of deionized, distilled water.
2. Mix in the following:
    a. 10 ml each Murashige and Skoog salts: nitrates, halides, NaFeEDTA, sulfates, and PBMo
    b. 10 ml thiamine stock (40 mg/liter)
    c. 10 ml myo-inositol stock (10 g/liter)
    d. 30 g sucrose
3. Adjust volume to 1000 ml. Adjust pH to 5.7.
4. Add 6 g TC agar. Melt.
5. Distribute 25 ml per culture tube (25 × 150 mm). Cap.
6. Autoclave for 15 min at 121°C, 15 psi.

Explant Preparation

Remove rhizome clumps from the Stage II medium and cut them into cubes measuring 1 cm on each side. Trim the fronds to a length of 15–20 mm. Place the plant material in culture tubes or jars containing Stage III medium. The number of plants placed in each container is determined by the size of the container being used. A Mason jar resting on its side can hold 50 plants, but a test tube can hold only 1 or 2. Push the rhizome end beneath the surface of the medium, leaving the fronds above the surface. The temperature and light requirements for this stage are the same as those for Stages I and II.

After the plant material has been recultured into the Stage III medium, rooting and plant development will usually be complete in 2–3 weeks, at which time the plants can be transferred to a potting mix. Roots will be clearly visible in the agar when the plants are ready for transplantation.

Transplantation

Whether transplanting shoots to a potting mix from Stage II or Stage III medium, the procedure is the same. Carefully wash the agar from the plants with tap water. If any agar clings to the ferns, it will encourage the growth of fungi, possibly killing the plants.

Examine the effects of different potting mixes on the establishment and growth of the transplanted ferns. Variance in growth will be observed. When the shoots are free of agar, they may be transplanted to a potting mix by making a hole in the mix with a sharp object and placing the root ball firmly into the potting mix. It is important that the roots have intimate contact with the mix and that the soil is firmly pressed around the stem. After transplantation, water the plants.

Potted plants must be protected for about 2 weeks or until the roots are visible in the holes at the bottom of the pots. During this time the pots should be draped with shade cloth or cheesecloth and placed near windows or under artificial lights at a light intensity of 500–1000 fc. Humidity should be kept high during this period by placing the pots on a moist bed of gravel or covering each pot with a plastic bag and securing the bag with a rubber band.

After 2 weeks begin to lower the humidity by loosening the plastic around the pots and allowing greater air circulation. After 8 weeks a full root system should be visible throughout the pot. Each fern can now be transferred to a larger pot or basket. Beginning 3 weeks after the ferns have undergone final transplantation, the plants should be fed weekly with a balanced fertilizer.

STAGHORN FERN

Sterile gametophyte tissue from aseptic spore germination is an excellent explant source to initiate fern cultures (Bashe, 1973; Beckwith & Schroder, 1978; Flifet, 1961; Knaus, 1976; Spear, 1977). Germination can occur within 7 days on an agar medium, and prothallia are evident within 14 days. Clumps rapidly increase in size and require division and subsequent transfer to media every 6 weeks. Aseptic germination is advantageous over the usual germination on a peat moss medium because it is less labor intensive.

Purpose: To initiate cultures of *Phatycerum bifurcatum* from spores and compare two techniques for large-scale multiplication (Hennen & Sheehan, 1978; Cooke, 1979).

Stage I

Medium Preparation: 500 ml liter equivalent, Spore Germination Medium.

1. Into a 1000-ml Erlenmeyer flask pour 250 ml of deionized, distilled water.
2. Mix in the following:
   a. 5 ml each Murashige and Skoog salts: nitrates, halides, NaFeEDTA, sulfates, and PBMo
   b. 5 ml thiamine stock (40 mg/liter)
3. Adjust volume to 500 ml. Adjust pH to 5.7.
4. Add 4 g TC agar.
5. Autoclave for 15 min at 121°C, 15 psi.
6. Distribute 25 ml per sterile plastic Petri dish (100 × 20 mm).

Explant Preparation

1. Collect spores from mature staghorn ferns by scraping spores from the underside of the fronds into tapered plastic centrifuge tubes.
2. Wash the spores by adding sterile water to each tube and centrifuging for 5 min; decant the supernatant.
3. Resuspend the spore pellet in 2% chlorine bleach with 2 drops Tween-20 per 100 ml. Mix for 10 min.
4. Centrifuge. Using an aseptic technique, decant the bleach solution and rinse three times in sterile water.
5. Resuspend the pellet of spores in 5 ml of sterile, distilled water.
6. Inoculate the medium with the spore suspension.
7. Culture in the dark for 1 week and then transfer to the illuminated culture shelf.

Observations

Prothallia development from the spores should be apparent by 4 weeks. Once fertilization has taken place and the sporophyte plant develops, the sporophyte plant can be divided and subcultured onto Stage II medium.

Stage II

Medium Preparation: 1 liter equivalent, Stage II Staghorn Fern Medium.

1. Into a 2000-ml Erlenmeyer flask pour 500 ml of deionized, distilled water.
2. Mix in the following:
   a. 10 ml each Murashige and Skoog salts; nitrates, halides, NaFeEDTA, sulfates, and PBMO
   b. 10 ml thiamine stock (40 mg/liter)
   c. 10 ml myo-inositol stock (10 g/liter)
   d. 30 g sucrose
3. Adjust volume to 1000 ml. Adjust pH to 5.7.

4. Add 8 g TC agar. Melt.
5. Distribute 25 ml per culture tube (25 × 150 mm). Cap.
6. Autoclave for 15 min at 121°C, 15 psi.
7. Cool as slants.

Explant Preparation

Divide Stage I cultures of sporophyte plant and culture on the culture shelf.

Observations

In about 6 weeks each culture should have multiplied into a mass of plants (20–40 plants). Sacrifice several cultures and separate them into individual plants; count the number of plants per culture. These cultures will be homogenized to initiate Stage III cultures (Cooke, 1979).

Stage III

Medium Preparation: 1 liter equivalent, Stage III Staghorn Fern Medium. Stage II Staghorn Fern Medium is used with the following steps 3–9:

3. Adjust volume to 1000 ml. Adjust pH to 5.7.
4. Divide into four 250-ml portions.
5. Add 2.25 g TC agar to each. Cap each flask with foil.
6. Autoclave for 15 min at 121°C, 15 psi.
7. When the medium has cooled to about 40°C, put the 250-ml portions into a sterile blender to homogenate in the transfer hood. Add the contents (separate plant material from agar—discard agar) of three cultures of Stage II staghorn ferns.
8. Blend for 5 s and pour 20 to 25-ml aliquots into sterile plastic Petri dishes (100 × 20 mm).
9. Place the Petri dishes on the culture shelf.

Observations

Growth should be evident in 3 weeks at all depths of the medium. In 2 months high rates of multiplication should be achieved. Each Petri dish should contain many plants. Count the number of plants per dish. The plants can easily be washed free of agar and established in a potting mix. The pots should be maintained under high humidity for 2 weeks. The survival rate should be about 80%.

Data and Questions

1. How many plants were you able to separate out from a Stage II culture? How long did this take?

2. What are the advantages of the homogenization technique?
3. About how many plant fragments did you end up with in your Petri dish? (Count the number on a quarter of the Petri dish, and multiply by four.) How many ferns developed, and how long did this take?

Plant shoot tip culture has proven itself to be one of the most widely used techniques for clonal propagation of ornamental plants. The method is labor intensive accounting for ~60% of the propagation cost. Attempts to mechanically excise and culture shoots via robotics and liquid culture systems have been limited (McCown, 2003). Woody plant multiplication in particular has been challenging (see Chapter 8). Kalinia and Brown (2007) reported micropropagation of nine ornamental *Prunus* spp. using shoot apex culture.

FICUS

Purpose: To perform shoot tip culture for axillary bud multiplication of *Ficus elastica "Decora."*

Stage I

Medium Preparation: 1 liter equivalent, Stage I Ficus Medium.

1. Into a 2000-ml Erlenmeyer flask pour 500 ml of deionized, distilled water.
2. Mix in the following:
   a. 7.5 ml each Murashige and Skoog salts: nitrates, halides, NaFeEDTA, sulfates, and PBMo
   b. 10 ml thiamine stock (10 g/liter)
   c. 10 ml myo-inositol stock (10 g/liter)
   d. 30 g sucrose
   e. 10 ml $NaH_2PO_4 \cdot H_2O$ stock (17 g/liter)
   f. 8 ml adenine sulfate stock (1 g/100 ml)
   g. 3 ml IAA stock (50 mg/500 stock) h. 300 ml 2iP stock (50 mg/500 ml)
3. Adjust volume to 1000 ml. Adjust pH to 5.7.
4. Add 8 g TC agar. Melt.
5. Dispense 25 ml per culture tube (25 × 150 mm). Cap.
6. Autoclave for 15 min at 121°C, 15 psi.

Explant Preparation

Remove 2- to 3-inch long shoot tips from stock plants. Wash the tips in warm, soapy water and disinfect in 10% chlorine bleach for 10 min; rinse three times. Aseptically remove the outer leaves and cut off bleached stem tissue. Culture the tips on the culture shelf. After 4–6 weeks the clean cultures can be moved into Stage II medium.

Stage II

Medium Preparation: 1 liter equivalent, Stage II Ficus Medium.

This medium is the same as the Stage I Ficus Medium except that IAA is omitted.

Stage III

Medium Preparation: 500 ml equivalent, Stage III Ficus Medium.

Into a 1000-ml Erlenmeyer flask pour 300 ml of deionized, distilled water.

1. Mix in the following:
   a. 5 ml each Murashige and Skoog salts: nitrates, halides, NaFeEDTA, sulfates, and PBMo
   b. 5 ml thiamine stock (40 mg/liter)
   c. 5 ml myo-inositol stock (10 g/liter)
   d. 15 g sucrose
   e. 4 ml adenine sulfate stock (1 g/100 ml)
   f. 2 ml IAA stock (10 mg/100 ml)
2. Adjust volume to 500 ml. Adjust pH to 5.7.
3. Add 4 g TC agar. Melt.
4. Dispense 25 ml per culture tube (25 × 150 mm). Cap.
5. Autoclave for 15 min at 121°C, 15 psi.

Explant Preparation

Separate individual shoots of Stage II cultures and culture in Stage III medium for rooting. When roots develop, the rooted shoot can be removed from culture, washed free of agar, and potted in a soil mix. Cover the plant to maintain humidity for approximately 2 weeks to harden off.

Questions

1. What is the difference between Stage I and Stage II regarding growth response from explant?
2. Is callus formed in the Stage II cultures?
3. What area of the explant gives rise to new shoots?
4. What is the potential rate of multiplication in Stage II?

KALANCHOE

Purpose: To observe regeneration of adventitious buds from shoot, stem, and leaf explants of kalanchoe and to observe the effect of genotypes on regeneration (Smith & Nightingale, 1979).

Medium Preparation: 1 liter equivalent, Kalanchoe Shoot Initiation Medium.

1. Into a 2000-ml Erlenmeyer flask pour 500 ml of deionized, distilled water.
2. Mix in the following:
   a. 10 ml each Murashige and Skoog salts: nitrates, halides, NaFeEDTA, sulfates, and PBMo
   b. 10 ml thiamine stock (40 mg/liter)
   c. 10 ml myo-inositol stock (10 g/liter)
   d. 30 g sucrose
   e. 10 ml kinetin stock (10 mg/100 ml)
   f. 8 ml adenine sulfate stock (17 g/liter)
   g. 10 ml $NaH_2PO_4 \cdot H_2O$ stock (17 g/liter)
   h. 10 ml IAA stock (10 mg/100 ml)
3. Adjust volume to 1000 ml. Adjust pH to 5.7.
4. Add 8 g TC agar to Difco-Bacto agar. Melt.
5. Distribute 25 ml per culture tube (25 × 150 mm). Cap.
6. Autoclave for 15 min at 121°C. 15 psi.

Explant Preparation

Leaves, stem sections, and shoot tips are collected from the *Kalanchoe bossfeldiana* Poelln. stock plant. The responses of different cultivars of kalanchoe will vary. Compare several cultivars. Wash leaf, stem, and shoot tip explants in warm, soapy water to remove dust and then rinse. Wrap the plant material in cheesecloth squares to prevent it from floating.

   Place the wrapped plant material in the container in which it is to be sterilized (culture tube or small beaker). To a 15% (v/v) bleach solution (15 ml chlorine bleach and 85 ml water) add 2 drops/100 ml of a wetting agent (Tween-20 or detergent). Pour this solution over the plant material. If the container does not have a cap, carry out this step in a transfer hood. Swirl or agitate the mixture for 10 min. In a transfer hood, pour off the chlorine bleach or remove the wrapped plant material with forceps and rinse three times in sterile water.

   Place the wrapped plant material in a sterile Petri dish. Unwrap and remove one piece of tissue at a time to a clean Petri dish. Cut off the bleached pieces of tissue (white areas).

   Place sections of leaf blade (15 × 15 mm), stem (15–20 mm), and shoot tips with two to four primordial leaves (1–2 mm) in tubes with nutrient medium. Incubate them on the culture shelf.

Questions

1. Record the appearance of adventitious buds from the different explants weekly. What is the average number of shoots from the different explants?
2. Which explant produces the most shoots?
3. From what area does an adventitious bud originate?

4. What differences between the responses of the different kalanchoe genotypes do you see?
5. Are there differences between the contamination rates for the different explants?

AFRICAN VIOLET

Purpose: To initiate clean cultures for rapid clonal propagation and to isolate various phenotypes from sections of chimeric African violet leaves and petioles (Bilkey *et al.*, 1978; Norris *et al.*, 1983; Start & Cummings, 1976).

Medium Preparation: 1 liter equivalent, African Violet Medium.

1. Into a 2000-ml Erlenmeyer flask pour 500 ml of deionized, distilled water.
2. Mix in the following:
   a. 10 ml each Murashige and Skoog salts: nitrates, halides, NaFeEDTA, sulfates, and PBMo
   b. 10 ml thiamine stock (40 mg/liter)
   c. 10 ml myo-inositol stock (10 g/liter)
   d. 30 g sucrose
   e. 20 ml of IAA stock (10 mg/100 ml)
   f. 20 ml kinetin stock (10 mg/100 ml)
   g. 10 ml $NaH_2PO_4 \cdot H_2O$ stock (17 g/liter) h. 8 ml adenine sulfate stock (1 g/100 ml)
3. Adjust volume to 1000 ml. Adjust pH to 5.7.
4. Add 8 g TC agar. Melt.
5. Distribute 25 ml per culture tube (25 × 150 mm). Cap.
6. Autoclave for 15 min at 121°C, 15 psi.
7. Cool as slants.

Explant Preparation

1. Remove healthy, young leaves from an African violet stock plants with a scalpel. Use different variegated cultivars with white, red, or light green sectoring of leaves. Record the variety name of each plant.
2. Wash the leaf explants in warm, soapy water and rinse. The leaf tissue can bruise, so be careful.
3. Wrap the explants in cheesecloth and place in a 250-ml beaker covered by a Petri dish.
4. Disinfect the tissue in 10% chlorine bleach for 15 min.
5. Rinse the tissue three times in sterile, distilled water.
6. Excise 1-cm^2 explants from the leaves and 1-cm sections of petiole. Each student should take six explants from the healthy leaves, as shown in Fig. 12.2. Cut the light sector from the leaf and include all of the main leaf vein adjacent to the mutant sector (even if the vein is green).

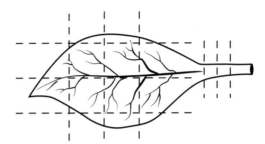

FIGURE 12.2   Excise six explants from the African violet leaf: four 1-cm² leaf blade sections and two 1-cm petiole sections. Cut the light sector from the leaf and include the main leaf vein adjacent to the mutant sector.

7. Place these explants on the African violet medium.
8. Incubate the explants on the culture shelf.

Data and Questions

Observe the weekly progress of the explants. Note differences in the development of the sections.

1. Is there a difference in morphogenetic response between leaf blade tissue and petiole tissue? If so, what is the difference?
2. What is a chimera? Note the location and color of plantlets arising from the different sectors.
3. What purpose do pure albino lines serve in tissue culture research?
4. Were there differences in response between the different cultivars. To what do you attribute them?

ENDANGERED SPECIES

Many plant populations of cacti, carnivorous plants, and endemic native populations worldwide are endangered because of declining habitat and extensive field collection. Some rare plants have been collected from their native habitats, thus endangering their survival. Conventional propagation of some cacti is not adequate to meet demand due to low offset numbers, poor seed germination, damping-off, sterility, and slow growth. About one-fourth of native cacti are rare. Many field-collected plants do not survive in cultivation. Micropropagation of these plants could halt trends toward extinction.

Reed *et al.* (2011) introduce an entire issue of *In Vitro Cell. Dev. Biol.-Plant* containing articles on conserving biodiversity and small remnant plant populations. Pence (2011) and Bunn *et al.* (2011) provide references and guidance for working with endangered plant populations. Additionally, extensive information on short- and long-term tissue storage and the cost for establishing, propagating, and preserving endangered species is discussed. Also included in the Bibliography are works discussing propagation techniques for specific endangered plants.

Pitcher Plants

Purpose: To establish multiplying cultures of the pitcher plant (Adams *et al.*, 1979).

    Medium Preparation: 1 liter equivalent, Pitcher Plant Multiplication Medium.

1. Into a 2000-ml Erlenmeyer flask pour 500 ml of deionized, distilled water.
2. Mix in the following:
   a. 2.5 ml each Murashige and Skoog salts: nitrates, halides, NaFeEDTA, sulfates, and PBMo
   b. 20 g sucrose
   c. 100 ml BA stock (10 mg/100 ml)
   d. 100 ml IBA stock (10 mg/100 ml)
3. Adjust volume to 1000 ml. Adjust pH to 5.7.
4. Add 8 g TC agar.
5. Autoclave for 15 min at 121°C, 15 psi.
6. Distribute 15 ml per sterile plastic Petri dish (60 × 15 mm).

Explant Preparation

Stock plants can be obtained from local nurseries. Place plastic drinking cups over the potted plants to maintain 100% humidity. The plants can be maintained under fluorescent lights at 20–25°C. Take 2-cm sections of leaf and stem plant material, disinfect in 20% chlorine bleach for 15 min, and rinse three times. Place cultures on the culture shelf. The pitcher plant contains endogenous microorganisms, making it difficult to establish Stage I cultures.

CACTUS

Purpose: To establish aseptic cactus seedlings of *Coryphantha* sp. or other cactus species (Ault & Blacknon, 1985, 1987; Clayton *et al.*, 1990; Johnson & Emino, 1979, 1979a; Mauseth, 1979; Smith *et al.*, 1991; Mariateresa *et al.*, 2010).

    Medium Preparation: 1 liter equivalent, Stage I Cactus Medium.

1. Into a 2000-ml Erlenmeyer flask pour 500 ml of deionized, distilled water.
2. Mix in 10 ml each Murashige and Skoog salts: nitrates, halides NaFeEDTA, sulfates, and PBMo
3. Adjust volume to 1000 ml. Adjust pH to 5.7.
4. Add 8 g TC agar. Melt.
5. Distribute 25 ml per culture tube (25 × 150 mm). Cap.
6. Autoclave for 15 min at 121°C, 15 psi.

Explant Preparation

Surface sterilize seeds in 10% chlorine bleach for 10 min and rinse in sterile water. Place the seeds on the surface of Stage I Cactus Medium. Culture on the culture shelf.

Shoot explants from the aseptically grown seedlings can be subcultured on MS medium supplemented with (per liter) 04 mg thiamine, 100 ml myo-inositol, 20 g sucrose, 44 µM BA, and 0.5 µM 2,4-D at pH 5.7. Callus will proliferate in 4 weeks and can be subcultured on the medium just described.

For shoot proliferation, subculture 10-mm diameter callus pieces on the preceding medium without the BA and 2,4-D. Multiple shoots will develop in 8 weeks. Shoots can be isolated and cultured on half-strength MS medium, 0.4 mg/liter thiamine, 20 g/liter sucrose, and 8 g/liter TC agar. The medium (10 ml per container) should be poured into Magenta GA-7 vessels. As the medium dries, rooting is enhanced, and by 8 weeks most shoots should be rooted.

BIBLIOGRAPHY

Adams, R. M., Koenigsberg, S. S., & Langhans, R. W. (1979). In vitro propagation of *Cephalotus follicularis (Australian pitcher plant)*. *HortScience, 14*, 512–513.

Adams, R. M., Koenigsber, S. S., & Langhans, R. W. (1979). In vitro propagation of the butterwort (*Pinguicula moranensis* H. B. K.). *HortScience, 14*, 701–702.

Arditti, J. (1977). Clonal propagation of orchids by means of tissue culture-A manual. In J. Arditti (Ed.), *Orchid biology, reviews and perspective* (1, pp. 203–293). Ithaca, NY: Cornell Univ. Press.

Ault, J. R., & Blackmon, W. J. (1985). In vitro propagation of selected native cacti species. *HortScience, 20*, 541(Abstr).

Ault, J. R., & Blackmon, W. J. (1987). In vitro propagation of *Ferocactus acanthodes (Cactaceae)*. *HortScience, 22*, 126–127.

Bashe, D. V. (1973). A simple method of initiating axenic cultures of pteridophytes from spores. *Amer. Fern J., 63*, 147–151.

Beckwith, W. J., & Schroder, L. C. (1978). Growing ferns from spores. *Southern Florist & Nurseryman, Dec., 28*, 26–27.

Benson, L. (1977). Preservation of cacti and management of the ecosystem. In G. T. Prance, & T. S. Elias (Eds.), *Extinction is forever* (pp. 283–300). New York: New York Botanical Garden.

Bilkey, P. C., McCown, B. H., & Hildebrandt, A. C. (1978). Micropropagation of African violet from petiole cross-sections. *HortScience, 13*, 37–38.

Bristow, J. M. (1962). The controlled *in vitro* differentiation of callus from a fern, *Pteris cretica* L., into gametophytic or sporophytic tissue. *Dev. Bio., 4*, 361–375.

Bunn, E., Turner, S. R., & Dixon, K. W. (2011). Biotechnology for saving rare and threatened flora in a biodiversity hotspot. *In Vitro Cellular & Dev. Biol.-Plant., 47*(1), 188–200.

Burr, R. W. (1976). Mass propagation of ferns through tissue culture. *In Vitro, 12*, 209–310.

Chen, S. Y., & Read, P. E. (1983). Micropropagation of Leatherleaf fern. *Proc. Fl. State Hort. Soc., 96*, 266–269.

Clayton, P. W., Hubstenberger, J. F., & Phillips, G. C. (1990). Micropropagation of members of the Cactaceae subtribe cactinae. *J. Amer. Hort. Soc., 115*, 337–343.

Cooke, R. C. (1977). The use of an agar substitute in the initial growth of Boston ferns in vitro. *HortScience, 12*, 339.

Cooke, R. C. (1979). Homogenization as an aid in tissue culture propagation of *Platycerium* and *Davallia. HortScience, 14,* 21–22.

Debergh, P. (1977). Symposium on tissue culture for horticultural purposes. *Acta Horticulturae, 78,* 6–9.

DeFossard, R. A. (1976). *Tissue culture for plant propagators.* Armidale, Australia: University of New England, Department of Continuing Education.

DeFossard, R. A. (1977). Tissue culture in horticulture—Perspective. *Acta Horticulturae, 78,* 450–455.

Evans, D. A., Sharp, W. R., & Flick, C. C. (1981). Plant regeneration from cell cultures. *Hort. Rev., 3,* 214–314.

Evenari, M. (1989). The history of research on white-green variegated plants. *Botany Review, 55,* 106–133.

Flifet, T. (1961). Growing ferns from spores. *Amer. Fern J., 51,* 113–127.

Gantait, S., Mandal, N., Bhattacharyya, S., & Kanti Das, P. (2010). An elite protocol for accelerated quality-cloning in *Gerbera jamesonii* Bolus cv. *Sciella. In Vitro Cell. Dev. Biol.-Plant, 46,* 537–548.

Garcia, R., Pacheco, G., Falcao, E., & Borges, G. (2011). Influence of type of explant, plant growth regulators, salt composition of basal medium and light on callogenesis and regeneration in *Passiflora suberosa* L. (Passifloraceae). *Plant Cell Tissue & Organ Culture, 106*(1), 47–54.

Goncalves, S., Escapa, A. L., Grevenstuk, T., & Romano, A. (2008). An efficient in vitro propagation protocol for *Pinguicula lusitanica,* a rare insectivorous plant. *Plant Cell, Tissue & Organ Culture, 95*(2), 239–243.

Gosal, S. S., Wani, S. H., & Kang, M. S. (2010). Biotechnology and crop improvement. *J. Crop Improvement, 24*(2), 153–217.

Harper, K., & Murashige, T. (1976). *Clonal multiplication of ferns in vitro.* Master's thesis, Riverside: University of California.

Hartmann, H. T., & Kester, D. E. (1975). *Plant propagation: principles and practices* (3rd ed.). Englewood Cliffs, NJ: Prentice-Hall.

Hennen, G. R., & Sheehan, T. J. (1978). *In vitro* propagation of *Platycerium stemaria* (Beauvois) Desv. *HortScience, 13,* 245.

Hires, C. S. (1940). Growing ferns from spores on sterile nutrient media. *J. N. Y. Bot. Gardens, 41,* 257–266.

Hirsch, A. M. (1975). The effect of sucrose on the differentiation of excised fern leaf tissue into either gametophytes or sporophytes. *Pl. Phy., 56,* 390–393.

Holdgate, D. P. (1977). Propagation of ornamentals by tissue culture. In J. Reinert, & Y. P. S. Bajaj (Eds.), *Applied and fundamental aspects of plant cell, tissue, and organ culture* (pp. 18–43). New York: Springer-Verlag.

Ivanova, M., & Van Staden, J. (2010). Natural ventilation effectively reduces hyperhydricity in shoot cultures of *Aloe plyphylla* Schonland ex Pillans. *Plant Growth Regulation, 60*(2), 143–150.

Iyyakkannu, S., Yeon, S. J., & Ryong, J. B. (2011). Micropropagation of *Hedera helix* L. "Mini". Prop. *Ornamental Plants, 11,* 125–130.

Johnson, J. L., & Emino, E. R. (1979). *In vitro* propagation of Mammilaria elongata. *HortScience, 14,* 605–606.

Johnson, J. L., & Emino, E. R. (1979a). Tissue culture propagation in the Cactaceae. *Cactus & Succulent J., 51,* 275–277.

Kalinia, A., & Brown, D. C. W. (2007). Micropropagation of ornamental *Prunus* spp. and GF 305 peach, a *Prunus* viral indicator. *Plant Cell Reports, 26*(7), 927–935.

Kevers, C., Franck, R., Strasser, R. J., Dommes, J., & Gaspar, T. (2004). Hyperhydricity of micro-propagated shoots: A typically stress-induced change of physiological state. *Plant Cell Tissue Organ Cult, 77*, 181–191.

King, R. M. (1957). Studies in the tissue culture of cacti. *Cactus & Succulent J., 29*, 102–104.

De Klerk, G.-J. (2002). Rooting of microcuttings: Theory and practice. *In Vitro, Plant, 38*, 415–422.

Knaus, J. F. (1976). A partial tissue culture method for pathogen-free propagation of selected ferns from spores. *Proc. Fl. State Hort. Soc., 89*, 363–365.

Krikorian, A. D. (1982). Cloning higher plants from aseptically cultured tissues and cells. *Bio. Rev., 57*, 151–218.

Kyte, L. (1987). *Plants from test tubes—An introduction to micropropagation.* Portland, OR: Timber Press.

Maheshwaramma, S., Reddy, D. L., & Reddy, S. S. (2008). Standardization of hormonal concentration and type of explants for the in vitro propagation of chrysanthemum cv. CO-1. *Prog. Research, 3*(2), 163–165.

Mariateresa, C., Daniela, B., & Giuseppe, C. (2010). In vitro propagation of *Obregonia denegrii* Fric. (Cactaceae). *Propagation of Ornamental Plants, 10*(1), 29–36.

Mauseth, J. D. (1977). Cactus tissue culture: A potential method of propagation. *Cactus & Succulent J., 49*, 80–81.

Mauseth, J. D. (1979). A new method for the propagation of cacti: Sterile culture of axillary buds. *Cactus & Succulent J., 51*, 186–187.

Mauseth, J. D. (1976). Cytokinin and gibberellic acid induced effects on the structure and metabolism of shoot apical meristem in *Opuntia polyacantha* (Cactaceae). *Amer. J. Bot., 63*, 1295–1301.

Mauseth, J. D. (1977). Cactus tissue culture: A potential method of propagation. *Cactus & Succulent J., 49*, 80–81.

Mauseth, J. D. (1982). A morphogenetic study of the ultrastructure of *Echinocereus engelmanii* (Cactaceae). IV. Leaf and spine primordia. *Amer. J. Bot., 69*, 546–550.

Mauseth, J. D., & Halperin, W. (1975). Hormonal control of organogenesis in *Opuntia polyacantha* (Cactaceae). *Amer. J. Bot., 62*, 869–877.

McCown, B. H. (2003). Biotechnology in horticulture: 100 years of application. *HortScience, 38*(5), 1026–1030.

Miller, L., & Murashige, T. (1976). Tissue culture propagation of tropical foliage plants. *In Vitro, 12*, 797–813.

Minocha, S. C., & Mehra, P. N. (1974). Nutritional and morphogenetic investigations on callus cultures of *Neomammillaria prolifera* Miller (Cactaceae). *Amer. J. Bot., 61*, 168–173.

Morel, G. (1964). Tissue culture—A new means of clonal propagation in orchids. *Am. Orchid Soc. Bull., 33*, 473–478.

Morel, G. (1966). Clonal propagation of orchids by meristem culture. *Cymbidium Soc. News, 20*, 3–11.

Murashige, T. (1974). Plant propagation through tissue cultures. *Ann. Rev. Pl. Physiol., 25*, 135–166.

Naylor, E. E., & Johnson, B. (1937). A histological study of vegetative reproduction in *Saintpaulia ionantha. Ame. J. Bot., 24*, 673–678.

Norris, R., Smith, R. H., & Vaughn, K. C. (1983). Plant chimeras used to establish *de novo* origin of shoots. *Science, 220*, 75–76.

Oinam, G., Yeung, E., & Kurepin, L. (2011). Adventitious root formation in ornamental plants: I. General overview and recent successes. *Propagation of Ornamental Plants, 11*(2), 78–90.

Oki, L. (1981). The modification of research procedures for commercial propagation of Boston ferns. In M. J. Constantin, R. R. Henke, K. W. Hughes, & B. V. Conger (Eds.), *Propagation of higher plants through tissue culture: Emerging technologies and strategies Proc. of an International Symposium, Knoxville, TN Envir. & Exper. Bot* (21, pp. 397–413). London Pergammon Press (imprint of Elsevier).

Oldfield, S. (1985). Whither international trade in plants? *New Scientist, 106*, 10–11.

Pedhya, M. A., & Mehta, A. R. (1982). Propagation of fern (*Nephrolepis*) through tissue culture. *Pl. Cell Rept., 1*, 261–263.

Pierik, R. L.M. (1979). *In vitro culture of higher plants*. Wageningen, The Netherlands: Ponsen en Looijen.

Pietropaolo, J., & Pietropaola, P. (1986). *Carnivorous plants of the world*. Portland, OR: Timber Press.

Pence, V. C. (2011). Evaluating costs for the in vitro propagation and preservation of endangered species. *In Vitro Cell. Dev. Biol.-Plant, 47*, 176–187.

Preece, J. E. (2010). Micropropagation in stationary liquid media. *Prop. of Ornamental Pl., 10*(4), 183–187.

Reed, B., Sarasan, V., Kane, M., Bunn, E., & Pence, V. C. (2011). Biodiversity conservation and conservation biotechnology tools. *In Vitro Cell. Dev. Biol.-Plant, 47*, 1–4.

Roberts, D. J. (1965). Modern propagation of ferns. *Proc. Internat. Pl. Prop. Soc., 15*, 317–321.

Rojas-Martinez, L., Visser, R. G. F., & deKlerk, G.-J. (2010). The hyperhydricity syndrome: Waterlogging of plant tissues as a major cause. *Propagation of Ornamental Plants, 10*(4), 169–175.

Rounsaville, T., Touchell, D., Ranney, T., & Blazich, F. A. (2011). Micropropagation of Mahonia "Soft Caress". *HortScience, 46*(7), 1010–1014.

Sasaki, K., Endo, M., & Inada, I. (2004). Effects of sampling season of explants and growth stage of mother plants on the regenerative capacity in cultured shoot apices of Chrysanthemum (*Dendranthema* X *grandiflorum* (Ramat.) Kitam.). *J. Japan. Soc. Hort. Sci., 73*(1), 36–41.

Schnell, D. E. (1976). *Carnivorous plants of the United States and Canada*. Winston-Salem, NC: John F. Blair.

Sivanesan, I., Song, J. Y., Hwang, S. J., & Jeong, B. R. (2011). Micropropagation of *Cotoneaster wilsonii* Nakai—A rare endemic ornamental plant. *Plant Cell. Tiss. Organ Cult, 105*, 55–63.

Skoog, F., & Miller, C. (1957). Chemical regulation of growth and organ formation in plant tissues cultured *in vitro. Symp. Soc. Expt. Biol., 11*, 118–131.

Skoog, F., & Tsui, C. (1948). Chemical control of growth and bud formation in tobacco stem segments and callus cultured *in vitro*. Amer. *J. Bot., 35*, 782–787.

Smith, R. H., & Nightingale, A. E. (1979). In vitro propagation of Kalanchoe. *HortScience, 14*, 20.

Smith, R. H., Burdick, P. J., Anthony, J., & Reilley, A. A. (1991). *In vitro* propagation of *Coryphantha macromeris* (Benson). *HortScience, 26*, 315.

Spear, E. J. (1977). Care of *Platycerium* sporelings. *Proc. Fl. State Hort. Soc., 90*, 128–129.

Start, N. D., & Cummings, B. G. (1976). *In vitro* propagation of *Saintpaulia ionantha* Wendl. *HortScience, 11*, 204–206.

Wetherell, D. F. (1982). *Introduction to in vitro propagation*. Wayne, NJ: Avery.

Yam, T. W., & Arditti, J. (2009). History of orchid propagation: A mirror of the history of biotechnology. *Plant Biotechnology Reports, 3*(1), 1–56.

Chapter 13

Protoplast Isolation and Fusion

Jungeun Kim Park, Sunghun Park and James E. Craven
Kansas State University

A protoplast is a plant cell from which the cell wall has been removed. After the rigid cell wall is removed, only a thin plasmalemma membrane surrounds the cell. Many uses of protoplasts have been explored (Horváth, 2009; Ko *et al.*, 2006; Tamaru *et al.*, 2002; Wu *et al.*, 2009). Unique hybrids that cannot be produced by sexual crossing can be developed by protoplast fusion of two different species such as potato and tomato (Melchers *et al.*, 1978; Power *et al.*, 1980; Cocking *et al.*, 1977; Gleba & Hoffman, 1980). Additionally, transformation of protoplasts with foreign genes can result in significant crop improvement (Rhodes *et al.*, 1988; Mazarei *et al.*, 2008). Direct DNA uptake can occur by electroporation or can be mediated by polyethylene glycol (PEG), resulting in transgenic plants (Cocking, 1977; Datta *et al.*, 1990; Fromm *et al.*, 1986; Rhodes *et al.*, 1988; Ren & Zhao, 2008). Protoplasts can also be useful in transient and stable expression assays where gene constructs are tested in a plant system (Yoo *et al.*, 2007; Zhang *et al.*, 2008; Zhang *et al.*, 2011).

Physiological problems of the cell may be studied by the use of protoplasts (Galun, 1981). Because no cell wall is present, the plasmalemma may be accessed easily, and its chemical structure (lipids, proteins, or enzymes) and physical properties (solute transport through the plasma membrane) may be

Plant Tissue Culture. Third Edition. DOI: 10.1016/B978-0-12-415920-4.00013-X

investigated. Organelles may be individually isolated from protoplasts, allowing for the examination of metabolite transport between different intracellular compartments. Examination of the effects of exogenously applied materials on cell activity is simplified by the use of protoplasts.

Cells used for protoplast isolation may come from various sources, such as callus, suspension cultures, and plant tissue. If plant tissue is used, young leaves are an excellent source of cells. When leaves are used, the epidermal layer of cells is removed, if possible, exposing the mesophyll to a prepared enzyme solution that allows digestion of the cell wall. If the epidermal layer cannot be easily removed, the leaves may be cut into small strips, exposing the mesophyll layer.

The cells walls are usually removed by enzymatic digestion. A macerating enzyme like pectinase is used to separate the tissue into individual cells. Cellulase or hemicellulase enzymes are used to digest the cell walls, leaving protoplasts.

When enzymatic digestion is used, it is important to keep the osmotic potential of the digesting medium at ideal conditions to prevent bursting or shriveling of the protoplasts. This is accomplished by mixing the enzymes in a solution of nutrient salts and vitamins and adjusting the osmotic potential with mannitol, sorbitol, or a combination of the two. The pH of the digesting medium must also be adjusted for optimum enzyme activity and cell growth.

PLANT PROTOPLAST ISOLATION AND FUSION

Purpose: To learn the principles and techniques of plant protoplast isolation and fusion.

Plant Tissue: *Gossypium hirsutum* (cotton) callus 14 days after transfer; brightly colored *Saintpaulia* sp. (African violet) petals; tobacco leaf tissue; *Nicotiana benthamiana* leaf tissue; *Quercus palustris* (oak) seedling leaves; rice seed callus; wheat callus; *Arabidopsis thaliana* leaf tissue; *Solanum lycopersicum* (tomato) leaves or fruit.

Supplies

1. Enzyme solutions (store at 4°C for up to 1.5 weeks)
   a. Cotton—0.5% pectinase in 0.7 M mannitol
   b. African violet—1% cellulase; 0.5% pectinase in 0.7 M mannitol
   c. Alternative—1% Onozuka Cellulase R-10; 0.4% Onozuka Macerozyme; 0.4 M mannitol; 40 mM MES-KOH (pH 5.6); 40 mM KCl; 20 mM $CaCl_2$
2. Two small and one large Petri dish
3. Two 30-ml beakers
4. 45–55 μm nylon mesh
5. Two round-bottom centrifuge tubes
6. Tabletop centrifuge
7. Pasteur pipets

8. 0.7 M mannitol
9. Two glass slides with cover slips
10. PEG solution
    a. 10 ml 0.7 M mannitol
    b. 5 g PEG
    c. 50 mM $CaCl_2 \cdot H_2O$
11. Evans blue dye in 0.7 M mannitol
12. Light microscope
13. W5 Solution (300 mM NaCl; 250 mM $CaCl_2$; 10 mM KCl; 10 mM glucose; 3 mM MES-KOH (pH 5.6))
14. MS storage solution (liquid MS media)
15. 1.5 ml microcentrifuge tubes
16. Chamber for vacuum infiltration
17. Aluminum foil
18. Disposable plastic hemocytometer.

Protoplast Liberation

Protoplast isolation can require long periods of time. For this reason, some of the steps should be done before the exercise.

An outline of the protoplast isolation schedule is shown in Fig. 13.1.

1. Pipe 5 ml of the cotton and African violet enzyme solutions into separate sterile plastic Petri dishes (60 × 15 mm)
2. Carefully tear the petals from two African violet flowers to expose the interior pigmented cells. Float the torn sections on the African violet enzyme solution. Cover the Petri dish with aluminum foil and place on shaker (40 rpm).
3. Transfer 500 mg of the cotton callus into the cotton enzyme solution; cover with foil and place on shaker (40 rpm).

Alternative Protocol

Optional: For recalcitrant tissue types—specifically tomato leaf—a brief sterilization greatly improves the yield of the protoplast isolation. One minute in 70% ethanol, 7 minutes in 20% Clorox.

1. Using a sterile surgical blade, slice the plant tissue into fine strips or squares. For very thin or weak tissue this can be done submerged in W5 or MS storage solution.
2. Place 0.5–1 g diced plant tissue into a sterile 1.5 ml microcentrifuge tube.
3. Fill the remaining space of the tube with alternative enzyme solution (about 1–2 ml).
4. With tube open to atmosphere, place into a vacuum infiltration chamber and evacuate for 30 min.

A

Cotton callus 500 mg	African violet petals
Incubate in 0.5% pectinase for 14 hr	Incubate in 0.5% pectinase for 14 hr
Filter through 45–55 µm nylon mesh screen	Filter through 45–55 µm nylon mesh screen
Centrifuge at 100 g for 5 min	Centrifuge at 100 g for 5 min
Decant supernatant, and resuspend pellet in 1 ml mannitol	Decant supernatant, and resuspend pellet in 1 ml mannitol
Centrifuge at 100 g for 5 min	Centrifuge at 100 g for 5 min
Decant supernatant, and resuspend pellet in 1 ml mannitol	Decant supernatant, and resuspend pellet in 1 ml mannitol

Stain with Evans blue dye   Stain with Evans blue dye

Place one drop of each suspension in same dish
Add PEG solution, and wait 15 min

B

◄ **Mesh**

FIGURE 13.1   Part A is an outline of the protoplast isolation schedule. The boxed-in steps have already been performed. The cotton callus filter is shown in Part B: 52-µm nylon mesh screen is placed around a 10-ml beaker with the bottom out.

5. Remove tube from vacuum, close lid, and cover with aluminum foil (or otherwise exclude light from the plant tissue). Incubate with moderate shaking for the indicated amount of time:
   a. Tomato leaves—3 hours
   b. Tobacco leaves, tomato fruit, rice callus, wheat callus, *Nicotiana benthamiana* leaves, *Arabidopsis thaliana* leaves—6 hours
   c. Oak seedling leaves—18–24 hours

Protoplast Purification

1. After 14 h filter the suspensions through 45- to 55-µm nylon mesh into 30-ml beakers.
2. Rinse the screens with 5 ml of 0.7 M mannitol.
3. Pour the suspensions into round-bottom centrifuge tubes and centrifuge at 100 g, for 5 min.
4. Decant the supernatant and gently resuspend the protoplast pellets in 10 ml of 0.7 M mannitol.
5. Centrifuge at 100 g for an additional 5 min, decant the supernatant, and resuspend the pellet in 1 ml of 0.7 M mannitol.

Alternative Protocol

1. After incubation, remove foil from tube and invert gently several times. Open the tube and agitate tissue *gently* with sterile forceps to help release protoplasts from leaf remnants.
2. Filter protoplast solution through fine mesh sieve. Wash leaf remnants left on mesh with 0.5 ml W5 solution.
3. Store protoplast solution in 1.5 ml microcentrifuge tube at 4°C overnight to allow the protoplasts to settle. A large green layer should appear at the bottom.
4. Remove the enzyme/W5 supernatant from the settled protoplast solution with a micropipette. Resuspend protoplasts in MS storage solution.

Vital Staining

1. Place one drop of each protoplast suspension on separate slides; do not put cover slips on slides.
2. Add one drop of Evans blue dye to each slide and observe immediately under the light microscope.
3. Describe the location of the dye. Are the protoplasts alive or dead? How do you know?

Protoplast Fusion

1. Place one drop (50 µl) of each protoplast suspension in the same plastic Petri dish. Gently shake the dish to ensure mixture of the suspensions.
2. Wait 5 min for the protoplasts to settle to the bottom.
3. Slowly add (one drop at a time) a total of 0.5 ml of the PEG solution, first to the periphery and then to the center of the mixed protoplast drops.
4. After 15 min, slowly add (one drop at a time) a total of 1 ml of 0.7 M mannitol solution to dilute the PEG solution.
5. Raise one end of the dish and wash the protoplast clumps adhering to the plastic with up to 9 ml of 0.7 M mannitol.
6. Remove the PEG and mannitol residue from the dish. Add a few drops of mannitol solution to the fused cells.
7. After 5 min observe the fused products on the inverted microscope.

Yield Determination

If information concerning the yield per gram plant tissue is desired, the mass of the incubation containment vessel (microcentrifuge tube or Petri dish) must be determined prior to incubation, as must the mass of the tissue to be digested and the mass of the tissue, enzyme solution, and containment vessel as a whole.

1. Measure the mass of the empty microcentrifuge tube (*after* marker labels have been applied).
2. Measure the mass of the microcentrifuge tube + plant tissue.
3. After incubation of plant tissue, filtering of protoplast solution, settling and resuspension in MS storage solution, obtain the mass of the tube and the protoplast solution together.
4. Carefully remove *exactly* 100 μ of protoplast solution from the tube using a micropipette. Obtain the new mass; the difference between this mass and that of step 3 is the mass of 100 uL of protoplast solution.
5. Apply 10 uL gently homogenized protoplast solution to a hemocytometer, per manufacturer instructions.
6. Under a light microscope, count the number of cells in each segment of the hemocytemeter's grid. Six replicate counts of different 0.25 mm × 0.25 mm squares should be sufficient. If possible, it is good practice to photograph each square chosen to be counted for later reference. Realize that this photograph cannot capture a truly accurate count, however, as the protoplasts will be located at different levels and will be thus impossible to capture with one image. When counting, this must be taken into account; vary the focus of the microscope within a certain range to obtain the most accurate count of protoplasts.
7. To obtain a value for the yield protoplasts per gram plant tissue, apply the following general calculation: Protoplasts/Gram Tissue = [Average # cells per square * Total mass of resuspended protoplast solution * 100 uL/(Volume of square * Mass of 100 uL of resuspended protoplast solution)]/Original mass of plant tissue.

Developing a Protoplast Isolation Protocol

While the procedures listed above may be applicable to a wide array of plant species and tissue types, they cannot cover every possibility. The isolation of protoplasts is a technique that must be tuned to each particular species and tissue upon which it is used for it to be truly effective. To that end, the following paragraph gives a brief guide to the development of the unique protocol necessary when dealing with a novel plant or tissue.

There are a number of factors which influence the success of a protoplast isolation. These include pH and osmotic potential of enzyme and storage solutions, length of incubation, temperature, containment of incubation, and the presence or absence of centrifugation. It is recommended that those wishing to develop their own protoplast isolation technique should use the procedures listed here as a baseline to conduct a series of experiments upon each of these variables, and thus determine which conditions are best suited to a particular novel isolation. Some plants may require additional steps; tomato, for example, for which it is necessary to subject the tissue to ethanol and Clorox briefly before digestion to obtain any significant protoplast yield. Others, such as oak, must be incubated for a much longer period of time. The protoplasts of some plants may be healthier when incubated in a Petri dish as opposed to a microcentrifuge tube; in others, the opposite may be the case. The particular conditions must be determined individually.

BIBLIOGRAPHY

Cassels, A. C., & Cocker, F. M. (1982). Seasonal and physiological aspects of the isolation of tobacco protoplasts. *Physiologia Plantarum, 56,* 69–79.

Cocking, E. C. (1960). A method for the isolation of plant protoplasts and vacuoles. *Nature, 187,* 927–929.

Cocking, E. C. (1972). Plant cell protoplasts—Isolation and development. *Annual Review of Plant Physiology, 23,* 29–50.

Cocking, E. C. (1977). Uptake of foreign genetic material by plant protoplasts. *International Review of Cytology, 48,* 323–343.

Cocking, E. C., George, D., Price-Jones, M. J., & Power, J. B. (1977). Selection procedures for the production of interspecies somatic hybrids of *Petunia hybrida* and *Petunia parodii. Plant Science Letters, 10,* 7–12.

Datta, S. K., Peterhans, A., Datta, K., & Potrykus, I. (1990). Genetically engineered fertile indica-rice recovered from protoplasts. *BioTechnology, 8,* 736–740.

Finer, J. J., & Smith, R. H. (1982). Isolation and culture of protoplasts from cotton (*Gossypium klotzschianum* Anderss.) callus cultures. *Plant Science Letters, 22,* 147–151.

Fromm, M. E., Taylor, L. P., & Walbot, V. (1986). Stable transformation of maize after gene transfer by electroporation. *Nature, 319,* 791–793.

Galun, E. (1981). Plant protoplasts as physiological tools. *Annual Review of Plant Physiology, 32,* 237–266.

Gleba, Y. Y., & Hoffman, F. (1980). *Arabidobrassica*: A novel plant obtained by protoplast fusion. *Planta, 149,* 112–117.

Horváth, E. (2009). Protoplast isolation from *Solanum lycopersiucum* L. leaf tissues and their response to short-term NaCl treatment. *Acta Biologica Szegediensis, 53,* 83–86.

Ko, J. M., Su, J., Lee, S., & Cha, H. C. (2006). Tobacco protoplast culture in a polydimethylsiloxane-based microfluidic cannel. *Protoplasma, 227,* 237–240.

Mazarei, M., Al-Ahmad, H., Rudis, M. R., & Stewart, N. C. (2008). Protoplast isolation and transient gene expression in switchgrass, *Panicum virgatum* L. *Biotechnology Journal, 3,* 354–359.

Melchers, G., Sacristan, M. D., & Holder, A. A. (1978). Somatic hybrid plants of potato and tomato regenerated from fused protoplasts. *Carlsberg Research Communications, 43,* 203–218.

Pilet, P. E. (Ed.), (1985). *The physiological properties of plant protoplasts.* New York: Springer-Verlag.

Power, J. B., Berry, S. F., Chapman, J. V., & Cocking, E. C. (1980). Somatic hybridization of sexually incompatible petunias: Petunia parodii, Petunia parviflora. *Theoretical and Applied Genetics, 57,* 1–4.

Ren, Y.-J., & Zhao, J. (2008). Optimization of electroporation parameters for immature embryos of indica rice (Oryza sativa). *Rice Science, 15,* 43–50.

Rhodes, C. A., Pierce, D. A., Mettler, D. M., Mascarenhas, D., & Detmer, J. J. (1988). Genetically transformed maize plants from protoplasts. *Science, 240,* 204–207.

Shepard, J. R., Bidney, D., & Shahin, E. (1980). Potato protoplasts in crop improvement. *Science, 208,* 17–24.

Tamaru, Y., Ui, S., Murashima, K., Kosugi, A., Chan, H., Doi, R. H., & Liu, B. (2002). Formation of protoplasts from cultured tobacco cells and *Arabidopsis thaliana* by the action of cellulosomes and pectate lyase from *Clostridium cellulovorans. Applied and Environmental Microbiology, 68,* 2614–2618.

Wu, F. H., Shen, S. C., Lee, L. Y., Lee, S. H., Chan, M. T., & Lin, S. C. (2009). Tape-*Arabidopsis* sandwich—A simpler *Arabidopsis* protoplast isolation method. *Plant Methods, 5,* 16.

Yoo, S. D., Cho, Y. H., & Sheen, J. (2007). *Arabidopsis* mesophyll protoplasts: A versatile cell system for transient gene expression analysis. *Nature Protocols, 2,* 1565–1572.

Zhang, W., Nilson, S. E., & Assmann, S. M. (2008). Isolation and whole-cell patch clamping of *Arabidopsis* guard cell protoplasts. *Cold Spring Harbor Protocols, 3,* 1–8.

Zhang, Y., Su, J., Duan, S., Ao, Y., Dai, J., Liu, J., Wang, P., Li, Y., Liu, B., Feng, D., Wang, J., & Wang, H. (2011). A highly efficient rice green tissue protoplast system for transient gene expression and studying light/chloroplast-related processes. *Plant Methods, 7,* 30.

Chapter 14

Agrobacterium-Mediated Transformation of Plants

Jungeun Kim Park, Sunghun Park, Qingyu Wu and Stuart Sprague
Kansas State University

There are three major approaches to inserting foreign genes into plants. These include direct DNA uptake by protoplasts by either electroporation or PEG-induced uptake, biolistics, or *Agrobacterium*-mediated gene transfer.

Protoplast uptake of foreign DNA has been described by Cocking (1960). This technology has resulted in transgenic corn and rice (Datta *et al.*, 1992; Rhodes *et al.*, 1988; Ren & Zhao, 2008; Zhang *et al.*, 2011). A problem with the

Plant Tissue Culture. Third Edition. DOI: 10.1016/B978-0-12-415920-4.00014-1
Copyright © 2013 Elsevier Inc. All rights reserved.
155

application of this technology is lack of routine cell culture methods to obtain plants from protoplasts of many important crop species.

Biolistics, or microprojectile bombardment, is another exciting technique. This method uses accelerated microcarriers (gold or tungsten particles) to penetrate and deliver DNA into the cell (Klein *et al.*, 1988). This particle delivery system has resulted in gene expression in many different plants and plant parts. Many important crop species have been stably transformed, including corn, rice, tobacco, sorghum, and soybeans (Assem & Hassan, 2008; Liu & Godwin, 2012; Lowe *et al.*, 2009; Rafiq *et al.*, 2006).

The following exercises demonstrate the third major method of inserting foreign genes into plants, *Agrobacterium*-mediated transformation (Ding *et al.*, 2009; Ibrahim *et al.*, 2010; Ozawa, 2009; Reyes *et al.*, 2010). *Agrobacterium* is a common soil bacterium that can invade wounded plant cells and insert foreign genes into the plant cell genome. *Agrobacterium*-mediated transformation of petunia leaf disk and shoot apex explants is examined. The leaf disk system was developed by Horsch *et al.* (1985) and represented a technological breakthrough allowing almost routine transfer of foreign genes into some dicotyledons—mainly those in the Solanaceae family (petunia, tobacco, and tomato). This method works for leaf explants that can regenerate by adventitious shoot formation. The leaf disk system overcame the problems inherent in the protoplast transformation systems (i.e., extended culture period and difficulty in regenerating plants from protoplasts). However, the leaf disk system has limitations. Few crop species can be regenerated from leaf disks, and adventitious shoot regeneration can result in somaclonal variation (Larkin & Scowcroft, 1981), which is not desirable in transformation. Fortunately, callus, cotyledon, hypocotyl, stem and seed tissues as well as somatic embryos and shoot tips can be target tissue for *Agrobacterium*-mediated transformation.

Initially, a limitation of the *Agrobacterium* transformation system was that it was felt to be limited to dicotyledons and some gymnosperms and not useful for monocotyledons (Potrykus, 1990). Recently, however, *Agrobacterium* has demonstrated an ability to transfer genes to monocotyledon tissue and produce transgenic plants (Park *et al.*, 1996, 2001; Chan *et al.*, 1993; He *et al.*, 2010; Hiei *et al.*, 1994; Peng *et al.*, 1995; Ranineri *et al.*, 1990; Smith & Hood, 1995; Wu *et al.*, 2007).

Along with the exercise on the leaf disk system, there is an exercise on the use of the shoot apex explant for transformation (Ulian *et al.*, 1988; Park *et al.*, 1996; Zapata *et al.*, 1998; Gould *et al.*, 1991). This approach can eliminate the problems of somaclonal variation and lack of ability to regenerate from leaf disks, protoplasts, or cell cultures.

The transformation exercises require the use of a laboratory that has met U.S. Department of Agriculture and university requirements and regulations for the handling of genetically engineered organisms, and they should not be conducted otherwise.

These experiments use *Agrobacterium tumefaciens* EHA 101, which contains the plasmids PGUS3, which has genes for kanamycin resistance (NPTII) and β-glucuronidase (GUS).

PETUNIA OR TOBACCO LEAF DISK

Sequence

1. Culture on Leaf Disk Preculture Medium for 2 days.
2. Inoculate leaf disk with *Agrobacterium* containing foreign genes for 15 s to 1 min.
3. Culture on fresh Leaf Disk Preculture Medium for 2 days.
4. Culture 2–4 weeks on Leaf Disk Antibiotic Shoot Proliferation Medium.
5. After 2 weeks excise individual shoots from cultures and place on Leaf Disk Antibiotic Rooting Medium.

Purpose: To obtain transgenic plants of *Petunia* sp. or *Nicotiana tabacum* by leaf disk cocultivation with *Agrobacterium tumefaciens* containing the foreign genes.

Preculture

Medium Preparation: 1 liter equivalent, Leaf Disk Preculture Medium.

1. Into a 2000-ml Erlenmeyer flask pour 500 ml of deionized, distilled water.
2. Mix in the following:
   a. 10 ml each Murashige and Skoog salts: nitrates, halides, NaFeEDTA, sulfates, and PBMo
   b. 10 ml thiamine stock (40 mg/liter)
   c. 10 ml myo-inositol stock (10 g/liter)
   d. 30 g sucrose
   e. 10 ml BA stock (10 mg/100 ml)
   f. 1 ml NAA stock (10 mg/100 ml)
   g. 10 ml vitamin stock
3. Adjust volume to 1000 ml. Adjust pH to 5.7.
4. Add 8 g TC agar or 2 g/liter Gelrite.
5. Autoclave for 15 min at 121°C, 15 psi.
6. Distribute 25 ml per sterile plastic Petri dish (100 × 20 mm).

Explant Preparation

Remove whole leaves from a petunia plant ("Rose Flash," Ball Seed Co.) and gently wash them in warm, soapy water. Sterilize the leaves in 15% (v/v) chlorine bleach for 10 min and rinse three to five times in sterile water. Use either a scalpel or an autoclaved paper punch to obtain leaf sections. Culture disks on Leaf Disk Preculture Medium for 2 days on the culture shelf. Leaf tissue from tobacco or petunia plants from aseptically germinated seeds is an excellent source of leaf disks and does not require surface disinfestation.

Cocultivation with *Agrobacterium*

1. Remove leaf sections after 2 days and cocultivate only the puffy leaf disks— the flat ones are not good.
2. Put them into a sterile Petri dish with *Agrobacterium* that was started in suspension culture 1–2 days previously.
3. Cocultivate 15 s to 1 min; be sure the leaf disks are covered, top and bottom, by *Agrobacterium* broth.
4. Blotting is very important. This is done in the Petri dish with three layers of dry filter paper. Lift the top filter paper off and place the leaf disks on top of the two layers of filter paper below. Cover the leaf disks with the top filter paper. Press gently to blot off excess *Agrobacterium*. A good inoculation turns the outermost 1 mm of the leaf disk margin a dark green. If the entire disk is dark green, there has been too much *Agrobacterium* infection, and the disk will die.
5. Reculture the disks on clean Leaf Disk Preculture Medium.
6. Place the cultures on the culture shelf for 2–3 days.

Shoot Transformation and Elimination of *Agrobacterium*

Antibiotic Stocks

1. Carbenicillin stock: Weigh out 500 mg carbenicillin and place in a 200-ml beaker; add 100 ml deionized, distilled water. Adjust pH to 5.7 and filter-sterilize before adding to the cooled medium.
2. Kanamycin stock: Weigh out 100 mg kanamycin and place in a 200-ml beaker; add 100 ml deionized, distilled water. Adjust pH to 5.7 and filter-sterilize before adding to the cooled medium.

Filter-sterilization can be accomplished by attaching a Millipore disposable filter (45 µm) on the outlet of a syringe (100-ml volume) and drawing the stock solution into the syringe. The antibiotic is filter-sterilized as it passes through the filter into the cooled medium. Follow the manufacture's instructions for using the filter.

Medium Preparation: 1 liter equivalent, Leaf Disk Antibiotic Shoot Proliferation Medium.

1. Into a 2000-ml Erlenmeyer flask pour 500 ml of deionized, distilled water.
2. Mix in the following:
   a. 10 ml each Murashige and Skoog salts: nitrates, halides, NaFeEDTA, sulfates, and PBMo
   b. 10 ml thiamine stock (40 mg/liter)
   c. 10 ml myo-inositol stock (10 g/liter)

 d. 30 g sucrose
 e. 10 ml vitamin stock
 f. 10 ml BA stock (10 mg/liter)
 g. 1 ml NAA stock (10 mg/liter)
3. Adjust volume to 800 ml. Adjust pH to 5.7.
4. Add 8 g TC agar.
5. Autoclave for 15 min at 121°C, 15 psi.
6. When the medium is cool (the flask containing the autoclaved medium can be placed in a water bath at 40°C to keep the medium from solidifying until the antibiotics are added), filter-sterilize the carbenicillin and kanamycin stock solutions as they are added to the medium; use a transfer hood. The kanamycin concentration can have a range of 100–3 g/liter. At 300 g/liter you can be more confident that the shoots you get are transformed, but you will not get many. If you lower the concentration to 200 or 100 g/liter you will get more shoots, some of which will not be transformed.
7. Mix thoroughly.
8. Pouring rapidly, distribute 10–20 ml per sterile plastic Petri dish (60 × 15 mm).

Procedure

Transfer the leaf disk cultures to the Leaf Disk Antibiotic Rooting Medium. When transferring the leaf disks to this medium, repeat the filter paper blotting to remove excess *Agrobacterium*. Shoots will be obvious in 2 weeks. At 2–4 weeks, transplant the shoots for rooting.

Rooting and Pretesting for Transformed Shoots

Medium Preparation: 1 liter equivalent, Leaf Disk Antibiotic Rooting Medium.

1. Into a 2000-ml Erlenmeyer flask pour 500 ml of deionized, distilled water.
2. Mix in the following:
 a. 10 ml each Murashige and Skoog salts: nitrates, halides, NaFeEDTA, sulfates, and PBMo
 b. 10 ml thiamine stock (40 mg/liter)
 c. 10 ml myo-inositol stock
 d. 30 g sucrose
 e. 10 ml vitamin stock
3. Adjust volume to 800 ml. Adjust pH to 5.7.
4. Add 8 g TC agar.
5. Autoclave for 15 min at 121°C, 15 psi.
6. When medium is cool, add 100 ml carbenicillin stock and 100 ml kanamycin stock; use a transfer hood.
7. Mix thoroughly.
8. Pouring rapidly, distribute 25 ml per sterile plastic Petri dish (100 × 20 mm).

Procedure

1. Excise shoots at 2–4 weeks and place individual shoots on Leaf Disk Antibiotic Rooting Medium.
2. Shoots that root on kanamycin are likely to be transformed.
3. A variation is to take leaf disks from rooted shoots and culture them on Leaf Disk Antibiotic Rooting Medium without carbenicillin. Callusing is an indication that the tissue is transformed.
4. Only molecular analysis of parent and progeny plants can provide true verification of transformation.

PETUNIA SHOOT APEX

Purpose: To obtain transgenic plants of Petunia sp. by shoot apex cocultivation with *Agrobacterium tumefaciens* containing the foreign genes.

Sequence

1. Aseptically germinate petunia seedlings.
2. Excise shoot apices (apical dome and two primordial leaves) from 1-week-old seedlings.
3. Culture for 2 days on Shoot Development Medium.
4. Inoculate (~15 min) with *Agrobacterium* and culture for 2 days on fresh Shoot Development Medium.
5. Culture 3–4 weeks on Shoot Apex Antibiotic Shoot Development Medium.
6. After 1–2 weeks root on Shoot Apex Antibiotic Rooting Medium.

Seed Germination

Medium Preparation: 1 liter equivalent, Seed Germination Medium.

1. Into a 2000-ml Erlenmeyer flask pour 500 ml of deionized, distilled water.
2. Mix in 10 ml each Murashige and Skoog salts: nitrates, halides, NaFeEDTA, sulfates, and PBMo.
3. Adjust volume to 1000 ml. Adjust pH to 5.7.
4. Add 8 g TC agar. Melt.
5. Distribute 50 ml per Magenta GA-7 container. Cap.
6. Autoclave for 15 min at 121°C, 15 psi.

Explant Preparation

Surface-sterilize petunia seeds in 10% chlorine bleach for 10 min and rinse three times in sterile water. This should be done in a capped conical centrifuge tube. Plant the seeds in a Magenta container and place the container on the culture shelf.

Shoot Induction

Medium Preparation: 1 liter equivalent, Shoot Development Medium.

1. Into a 2000-ml Erlenmeyer flask pour 500 ml of deionized, distilled water.
2. Mix in the following:
    a. 10 ml each Murashige and Skoog salts: nitrates, halides, NaFeEDTA, sulfates, and PBMo
    b. 10 ml thiamine stock (40 mg/liter)
    c. 10 ml myo-inositol stock (10 g/liter)
    d. 30 g sucrose
    e. 1 ml BA stock (10 mg/100 ml)
3. Adjust volume to 1000 ml. Adjust pH to 5.7.
4. Add 8 g TC agar.
5. Autoclave for 15 min at 121°C, 15 psi.
6. Distribute 25 ml per sterile plastic Petri dish (100 × 20 mm).

Explant Preparation

Under a dissecting microscope in a hood, excise the apical dome with two primordial leaves from 1-week-old seedlings. Culture for 2 days on Shoot Development Medium.

Shoot Apex Transformation

Medium Preparation: 1 liter equivalent, Shoot Apex Antibiotic Shoot Development Medium.

Follow steps 1 and 2 of the preceding procedure for Shoot Development Medium and then do the following:

3. Adjust volume to 800 ml. Adjust pH to 5.7.
4. Add 8 g TC agar.
5. Autoclave in an Erlenmeyer flask for 15 min at 121°C, 15 psi.
6. When the medium is cool, add filter-sterilized antibiotic stocks: 100 ml carbenicillin stock and 100 ml kanamycin stock.
7. Mix medium and distribute 25 ml per sterile plastic Petri dish (100 × 25 mm).

Rooting

Medium Preparation: 1 liter equivalent, Shoot Apex Antibiotic Rooting Medium.

Follow the procedure for Shoot Apex Antibiotic Shoot Development Medium, using MS salts, 3% sucrose, 100 mg/liter kanamycin, and 500 mg/liter carbenicillin. Root shoots in culture tubes (25 × 150 mm).

Observations

Observe the number of shoots that develop and root on the antibiotic medium. Growth on antibiotic medium is only a preliminary indication of transformation.

TOBACCO LEAF INFILTRATION

Purpose: Transient expression of desired proteins in tobacco leaves with *Agrobacterium tumefaciens* containing the foreign genes for fast analysis of protein subcelluar location or biochemical characters.

Sequence (see Fig. 14.1)

1. Grow *Agrobacterium* in 5 ml YEP medium for 1 day.
2. Harvest the *Agrobacterium* and resuspend with infiltration medium.
3. Culture on fresh Leaf Disk Preculture Medium for 2 days.
4. Infiltrate tobacco leaves and allow tobacco plants to grow under normal growth conditions for 2–7 days to express the desired proteins.
5. Biochemical or microscopy analysis should be conducted every 24 h between 2–7 days after infiltration to get the best results.

Grow *Agrobacterium*

YEP Medium Preparation: 1 liter equivalent, YEP medium.

1. Into a 1000-ml medium bottle pour 800 ml of deionized, distilled water.
2. Mix in the following:
   a. 10 g yeast extract
   b. 10 g trytone

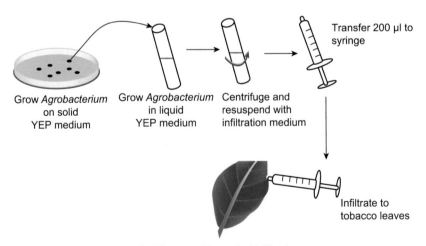

FIGURE 14.1   Tobacco leaf infiltration.

c. 5 g NaCl
d. 30 g sucrose
e. Add 15 g agarose if making solid YEP medium.
3. Adjust volume to 1000 ml.
4. Autoclave for 15 min at 121°C, 15 psi.
5. For solid medium, cool down in 60°C water bath and then add appropriate antibiotics. After evenly mixing the antibiotics with the medium, distribute 25 ml per sterile plastic Petri dish (100 × 20 mm).
6. For liquid medium, after autoclave, store in 4°C water before using.

Culture *Agrobacterium*

1. Streak the *Agrobacterium* stock on solid medium with appropriate antibiotics and culture under 28°C for 2 days.
2. Pick up single colony and inoculate with 5 ml YEP liquid medium with appropriate antibiotics; grow at 28°C for 1 day on 200-rpm shaker.

Harvest *Agrobacterium*

Infiltration Medium Preparation: 50 ml equivalent.

1. 10 ml 50 mM $MgCl_2$.
2. 5 ml 0.5 mM MES-KOH pH 5.6.
3. 5 μ 1 M acetosyringone.
4. Add 35 ml distilled water to 50 ml.

Procedure

1. Pellet cells at 1000 g, 10 min at 4°C, remove supernatant, add 5 ml of infiltration media and resuspend. Repeat again to remove the antibiotics, which will kill the leaf tissue after infiltration.
2. Adjust OD 600 of *Agrobacterium* to 1.0.
3. Take up 200 μl resuspended *Agrobacterium* in a 1 ml syringe (no needle).

Infiltration Tobacco Leaves

1. Remove 6-week-old, healthy tobacco plants (both *Nicotiana tabacum* and *N. benthamiana* work fine) from the growth chamber.
2. Choose healthy and fully expanded leaves for infiltration. Carefully mark the regions that will be infiltrated using the marker pen.
3. Place the tip of the syringe against the abaxial surface of the leaf in the marked region and press down gently on the plunger while directly supporting the adaxial side of the leaf with your finger.
4. Clean up the extra *Agrobacterium* left on the surface of leaves and pots.

5. Return plants in growth chamber and culture under normal growth conditions.
6. The desired proteins should be properly expressed after 2 days of growth, and need to be tracked every 24 hours.

BIBLIOGRAPHY

Assem, S. K., & Hassan, O. S. (2008). Real time quantitative PCR analysis of transgenic maize plants produced by *Agrobacterium*-mediated transformation and particle bombardment. *Journal of Applied Sciences Research, 4*, 408–414.

Caplan, A., Herrera-Estrella, L., Inze, D., Van Haute, E., Van Montagu, M., Schell, J., & Zambryski, P. (1983). Introduction of genetic material into plant cells. *Science, 222*, 815–821.

Chan, M. T., Chang, H. H., Ho, S. L., Tong, W. F., & Yu, S. M. (1993). *Agrobacterium*-mediated production of transgenic rice plants expressing a chimeric α-amylase promoter/β-glucuronidase gene. *Plant Molecular Biology, 22*, 491–506.

Chilton, M. D., Drummond, M. H., Merlo, D. J., Sciaky, D., Montoya, A. L., Gordon, M. P., & Nester, N. W. (1977). Stable incorporation of plasmid DNA into higher plant cells: The molecular basis of crown gall tumorigenesis. *Cell, 11*, 263–271.

Christou, P., Swain, W. F., Yang, N. S., & McCabe, D. E. (1989). Inheritance and expression of foreign genes in transgenic soybean plants. *Proceedings of the National Academy of Sciences, USA, 86*, 7500–7504.

Daniell, H., Vivekananda, J., Nielsen, B. L., Ye, G. N., Tewari, K. K., & Sanford, J. C. (1990). Transient foreign gene expression in chloroplasts of cultured tobacco cells after biolistic delivery of chloroplast vectors. *Proceedings of the National Academy of Sciences, USA, 87*, 88–92.

Datta, S. K., Datta, K., Soltanifar, S., Donn, G., & Potrykus, I. (1992). Herbicide-resistant Indica rice plants from IRRI breeding line IR72 after PEG-mediated transformation of protoplasts. *Plant Molecular Biology, 20*, 619–629.

Depicker, A., Van Montagu, M., & Schell, J. (1982). Plant cell transformation by Agrobacterium plasmids. In T. Kosuge, & C. Meredith (Eds.), *Genetic engineering of plants* (pp. 143–176). New York: Plenum.

Ding, L., Li, S., Gao, J., Wang, Y., Yang, G., & He, G. (2009). Optimization of Agrobacterium-mediated transformation conditions in mature embryos of elite wheat. *Molecular Biology Reports, 36*, 29–36.

Fraley, R. T., Horsch, R. B., Matzke, A., Chilton, M. D., Chilton, W. S., & Sanders, P. R. (1984). In vitro transformation of Petunia cells by an improved method of co-cultivation with A. tumefaciens strains. *Plant Molecular Biology, 3*, 371–378.

Fraley, R. T., Rogers, S. G., & Horsch, R. B. (1986). Genetic transformation in higher plants. *Critical Reviews of Plant Science, 4*, 1–46.

Gordon-Kamm, W. J., Spencer, M. T., Mangano, M. L., Adams, T. R., Daines, R. J., Start, W. G., O'Brien, J. V., Chambers, S. A., Adams, W. R., Jr., Willetts, N. G., Rice, T. B., Mackey, C. J., Krueger, R. W., Kausch, A. P., & Lemaux, P. G. (1990). Transformation of maize cells and regeneration of fertile transgenic plants. *Plant Cell, 2*, 603–618.

Gould, J., Devey, M., Ulian, E. C., Hasegawa, O., Peterson, G., & Smith, R. H. (1991). Transformation of Zea mays L. using *Agrobacterium tumefaciens* and the shoot apex. *Plant Physiology, 95*, 426–434.

He, Y., Jones, H. D., Chen, S., Chen, X. M., Wang, D. W., Li, K. X., Wang, D. S., & Xia, L. Q. (2010). *Agrobacterium*-mediated transformation of durum wheat (*Triticum turgidum* L. var. durum cv Stewart) with improved efficiency. *Journal of Experimental Botany, 61*, 1567–1581.

Hiei, Y., Ohta, S., Komari, T., & Kumashire, T. (1994). Efficient transformation of rice (Oryza sativa L.) mediated by Agrobacterium and sequence analysis of the boundaries of the T-DNA. *Plant Journal, 6*, 271–282.

Hinchee, M. A. W., & Horsch, R. B. (1986). Cellular components involved in the regeneration of transgenic Petunia hybrida. In D. A. Somer, B. G. Genenbach, D. D. Biesber, W. P. Hackett, & C. E. Green (Eds.), *VIth international congress of plant tissue and cell culture* (pp. 129). Minneapolis: University of Minnesota Press.

Horsch, R. B., Fry, J. E., Hoffman, N. L., Eichhlotz, D., Rogers, S. G., & Fraley, R. T. (1985). A simple and general method for transferring genes into plants. *Science, 227*, 1229–1231.

Ibrahim, A. S., El-Shihy, O. M., & Fahmy, A. H. (2010). Highly efficient *Agrobacterium tumefaciens*-mediated transformation of elite Egyptian barley cultivars. *American–Eurasian Journal of Sustainable Agriculture, 4*, 403–413.

Jefferson, R. A. (1987). Assaying chimeric genes in plants: The GUS gene fusion system. *Plant Molecular Biology Reports, 5*, 387–405.

Jefferson, R. A., Burgess, S. M., & Hirsch, D. (1986). Glucuronidase from Escherichia coli as a gene-fusion marker. *Proceedings of the National Academy of Sciences, USA, 83*, 8447–8451.

Jefferson, R. A., Klass, M., Wolf, N., & Hirsch, D. (1987). Expression of chimeric genes in Caenorhabditis elegans. *Journal of Molecular Biology, 193*, 41–46.

Kartha, K. K., Chibba, R. N., Georges, F., Leung, N., Caswell, K., Kendall, K., & Qureshi, J. (1989). Transient expression of chloramphenicol acetyltransferase (CAT) gene in barley cell cultures and immature embryos through microprojectile bombardment. *Plant Cell Reports, 8*, 429–432.

Klee, H., Horsch, R., & Rogers, S. (1987). Agrobacterium-mediated plant transformation and its further applications to plant biology. *Annual Review of Plant Physiology, 38*, 467–481.

Klein, T. M., Fromm, M., Weissinger, A., Tomes, D., Schaaf, S., Sletten, M., & Sanford, J. C. (1988). Transfer of foreign genes into intact maize cells with high-velocity microprojectiles. *Proceedings of the National Academy of Sciences, 85*, 4305–4309.

Klein, T. M., Kornstein, L., Sanford, J. C., & Fromm, M. E. (1989). Genetic transformation of maize cells by particle bombardment. *Plant Physiology, 91*, 440–444.

Larkin, P. J., & Scowcroft, W. R. (1981). Somaclonal variation—Novel source of variability from cell cultures for plant improvement. *Theoretical and Applied Genetics, 60*, 197–214.

Li, J. F., Park, E., Arnim, A. G., & Nebenfuhu, A. (2009). The FAST technique: A simplified Agrobacterium-based transformation method for transient gene expression analysis in seedlings of Arabidopsis and other plant species. *Plant Methods, 5*, 6–21.

Liu, G., & Godwin, I. (2012). Highly efficient sorghum transformation. *Plant Cell Reports, 31*, 1–9.

Lowe, B. A., Prakash, N. S., Way, M., Mann, M. T., Spencer, T. M., & Boddupalli, R. S. (2009). Enhanced single copy integration events in corn via particle bombardment using low quantities of DNA. *Transgenic Research, 18*, 831–840.

Morel, G. (1972). Morphogenesis of stem apical meristem cultivated in vitro: Applications to clonal propagation. *Phytomorphology, 22*, 265–277.

Morikawa, H., Iida, A., & Yamada, Y. (1989). Transient expression of foreign genes in plant cell and tissues obtained by a simple biolistic device (particle-gun). *Applied Micro Biotechnology, 31*, 320–322.

Ozawa, K. (2009). Establishment of a high efficiency *Agrobacterium*-mediated transformation system of rice (Oryza sativa L.). *Plant Science, 176*, 522–527.

Park, S. H., Pinson, S. R. M., & Smith, R. H. (1996). T-DNA integration into genomic DNA of rice following Agrobacterium inoculation of isolated shoot apices. *Plant Molecular Biology, 32*, 1135–1148.

Park, S. H., Park, J., & Smith, R. H. (2001). Herbicide and insect resistant elite transgenic rice. *Journal of Plant Physiology, 158*, 1221–1226.

Potrykus, I. (1990). Gene transfer to cereals: An assessment. *Bio/Technology, 8*, 535–542.

Rafiq, M., Fatima, T., Husnain, T., Bashir, K., Khan, M. A., & Riazuddin, S. (2006). Regeneration and transformation of an elite inbred line of maize (*Zea mays* L.), with a gene from Bacillus thuringiensis. *South African Journal of Botany, 72*, 460–466.

Raineri, D. M., Bottino, P., Gordon, M. P., & Nester, E. W. (1990). Agrobacterium-mediated transformation of rice (Oryza sativa L.). *Bio/Technology, 8*, 33–38.

Ren, Y.-J., & Zhao, J. (2008). Optimization of electroporation parameters for immature embryos of indica rice (oryza sativa). *Rice Science, 15*, 43–50.

Reyes, F. C., Sun, B., Guo, H., Gruis, D., & Otegui, M. S. (2010). Agrobacterium tumefaciens-mediated transformation of maize endosperm as a tool to study endosperm cell biology. *Plant Physiology, 153*, 624–631.

Rhodes, C. A., Pierce, D. A., Mettler, I. J., Mascernhas, D., & Detmer, J. J. (1988). Genetically transformed maize plants from protoplasts. *Science, 240*, 204–207.

Schrammeijer, B., Sijmons, P. C., van den Elzen, P. J. M., & Koekema, A. (1990). Meristem transformation of sunflower via Agrobacterium. *Plant Cell Reports, 9*, 55–60.

Simpson, R. B., Spielman, A., Margossian, L., & McKnight, T. D. (1986). A disarmed binary vector from Agrobacterium tumefaciens functions in Agrobacterium rhizogenes. *Plant Molecular Biology, 6*, 403–415.

Smith, R. H., & Hood, E. E. (1995). Agrobacterium tumefaciens transformation of monocotyledons. *Crop Science, 35*, 301–309.

Sparkes, I. A., Runions, J., Kearns, A., & Hawes, C. (2006). Rapid, transient expression of fluorescent fusion proteins in tobacco plants and generation of stably transformed plants. *Nature Protocol, 1*, 2019–2025.

Trinh, T. H., Mante, S., Pua, E. C., & Chua, N. H. (1987). Rapid production of transgenic flowering shoots and F1 progeny from Nicotiana plumbaginifolia epidermal peels. *Biotechnology, 5*, 1081–1084.

Ulian, E. C., Smith, R. H., Gould, J. H., & McKnight, T. D. (1988). Transformation of plants via the shoot apex. *In Vitro Cellular & Developmental Biology, 21*, 951–954.

Wang, Y. C., Klein, T. M., Fromm, E., Cao, J., Sanford, J. C., & Wu, R. (1988). Transient expression of foreign genes in rice, wheat and soybean cell following particle bombardment. *Plant Molecular Biology, 11*, 433–439.

Wu, H., Doherty, A., & Jones, H. D. (2007). Efficient and rapid Agrobacterium-mediated genetic transformation of durum wheat (Triticum turgidum L. var. durum) using additional virulence genes. *Transgenic Research, 17*, 425–436.

Zapata, C., Park, S. H., El-Zik, K. M., & Smith, R. H. (1999). Transformation of a Texas cotton cultivar by using Agrobacterium and the shoot apex. *Theoretical & Applied Genetics, 98*, 252–256.

Zhang, Y., Su, J., Duan, S., Ao, Y., Dai, J., Liu, J., Wang, P., Li, Y., Liu, B., Feng, D., Wang, J., & Wang, H. (2011). A highly efficient rice green tissue protoplast system for transient gene expression and studying light/chloroplast-related processes. *Plant Methods, 7*, 30–43.

Useful Measurements

LENGTH

1 inch = 25.4 millimeter (mm) = 2.54 centimeter (cm)
1 mm = 0.03937 inch

TEMPERATURE

32°F = 0°C	50°F = 10°C
68°F = 20°C	80°F = 27°C
86°F = 30°C	212°F = 100°C

CONCENTRATION

$$1 \text{ part per million (ppm)} = \frac{1 \text{ mg}}{1 \text{ kg}} = \frac{1 \text{ mg}}{1 \text{ liter of } H_2O}$$

$$\text{number of moles} = \frac{\text{weight in grams}}{\text{molecular weight}}$$

$$1 \text{ molar solution} = 1\,M = \frac{1 \text{ mole of solute}}{1 \text{ liter of solvent}}$$

$$1 \text{ millimolar solution} = mM = \frac{0.001 \text{ mole of solute}}{1 \text{ liter of solvent}}$$

$$1 \text{ micromolar solution} = 1\,\mu M = 0.001 \text{ m}M = \frac{1 \times 10^{-6} \text{ mole of solute}}{1 \text{ liter of solvent}}$$

WEIGHT

$$1 \text{ gram (g or gm)} = \frac{1}{453}\text{lb}$$

$$1 \text{ milligram (mg)} = \frac{1}{1000}\text{g}$$

$$1 \text{ microgram (μg)} = \frac{1}{1000}\text{mg} = 1 \times 10^{-6}\text{g}$$

Review of Solution Preparation

PERCENTAGE SOLUTIONS

In published papers the concentrations of sucrose, chlorine bleach, and agar are usually expressed as percentages. The concentration of the solute (agar, chlorine bleach, sucrose, etc.) is specified in 100 ml of a solution. This can be expressed as a weight per volume (w/v) or as a volume per volume (v/v).

Examples

A 10% (v/v) chlorine bleach solution is 10 ml of chlorine bleach diluted with water to a total volume of 100 ml. An 8% (w/v) agar concentration is 8 g of agar diluted to a total volume of 100 ml.

Problems

1. If a culture medium requires 4.5% (w/v) sucrose, how much sucrose would you add to make the following amounts of medium?
   a. 1 liter
   b. 500 ml
   c. 250 ml
   d. 10 ml
2. A 1% (w/v) agar concentration would require _____ g of agar in 1 liter of medium and _____ mg of agar in 20 ml of medium.
3. There are 7.5 g of agar in 250 ml of medium. What is the percentage of agar?
4. If 5 ml of chlorine bleach is diluted to 50 ml, what is the percentage of the chlorine bleach solution?

Answers

1. a. 45 g b. 22.5 g c. 11.2 g d. 450 mg
2. 10 g; 200 mg
3. 3%
4. 10%

MOLAR SOLUTIONS

A 1 M solution contains 1 mole of the solute in 1 liter of the solution. A mole is the number of grams of the solute that equals its molecular weight.

Problems

1. A 0.01 M solution of IAA (MW = 175) contains _____ g IAA per liter.
2. In a salt stress study 17 mM NaCl (MW = 58) was added to the medium to induce osmotic stress.
   a. How many grams of NaCl are in 1 liter of 17 mM NaCl solution?
   b. What is the percentage (w/v) of NaCl in 1 liter?
3. If you want 0.1 M IBA (MW = 203) in 1 liter of medium, how much IBA, in grams and milligrams, will you need?

Answers

1. 1 mole = 175 g
   175 g/liter = 1 M
   17.5 g/liter = 0.1 M
   1.75 g/liter = 0.01 M
2a. 17 mM = 0.017 M
   0.017 M = X g NaCl/58 MW
   0.017 × 58 = X g NaCl
   0.986 g NaCl/liter
2b. 0.986 g/1000 ml 0.0986 g/100 ml = 0.0986 or 0.1%
3. 203 g/liter = 1 M
   20.3 g/liter = 0.1 M = 20,300 mg/liter

DILUTION OF A STOCK SOLUTION

Problems

1. If you need 2.5 mg/liter, 2,4-D in 1 liter of medium and the stock solution of 2,4-D contains 10 mg/100 ml, how much stock solution do you need?
2. If you have a 15 mg/100 ml stock solution of GA$_3$ and you need a 1 mg GA$_3$ in 25 ml, how much stock solution would you add to 125 ml of medium?
3. There are 2150 mg kinetin (MW = 215) in 1 liter of medium. Express this in terms of molarity.

Answers

1. $10 \text{ mg}/10 \text{ ml} = 2.5 \text{ mg}/X$
   $10 \text{ mg } X = 250 \text{ mg/ml}$
   $X = 25 \text{ ml}$
2. $1 \text{ mg}/25 \text{ ml} = X/125 \text{ ml}$
   $25 \text{ ml } X = 125 \text{ mg/ml}$
   $X = 5 \text{ mg GA}_3 \text{ in } 125 \text{ ml medium}$
   $15 \text{ mg}/100 \text{ ml} = 5 \text{ mg}/X$
   $15 \text{ mg } X = 500 \text{ ml/mg}$
   $X = 33.3 \text{ ml stock solution}$
3. $2150 \text{ mg} = 2.15 \text{ g}$
   $2.15 \text{ g}/215 \text{ MW} = 0.01 \text{ moles}$
   $0.01 \text{ moles/liter} = 0.01 \ M$

List of Suppliers

Many scientific supply companies sell plant cell culture reagents, plant cell culture media, antibiotics, glassware, equipment, classroom kits, media supplements, and plasticware. Some of these companies are listed below.

BellCo Glass, Inc., www.bellcoglass.com
Bio-World, www.bio-world.com
Caisson Laboratories, Inc., www.caissonlabs.com
Carolina Biological Supply Company, www.carolina.com
Duchefa Biochemie, www.duchefodirect.com
Fisher Scientific, www.fishersci.com
Flow Laboratories, Inc., www.germfree.com
Hoechst Celanese Corporation, 1041 Route 202–206, Bridgewater, New Jersey 08807, 908.231.2000
Inotech Biosystems International, Inc., www.inotechintl.com
Li-Cor Biosciences, www.licor.com
Life Technologies, Inc., www.lifetechnologies.com
Millipore Corporation, www.millipore.com
Osmotek LTD, www.osmotek.com
Phenix Research, phenixresearch.com
PhytoTechnology Laboratories, www.phytotechlab.com
Plant Cell Technology, Inc., www.ppm4plant-tc.com
Sigma-Aldrich Company, www.sigmaaldrich.com
VWR Scientific, www.vwrsp.com

Common Plant Tissue Culture Terms[1]

Adventitious Developing from unusual points of origin, such as shoots or root tissues from callus, or embryos from sources other than zygotes.

Aneuploid The state of a cell nucleus that does not contain an exact multiple of the haploid number of chromosomes, one or more chromosomes being present in greater or lesser number than the rest. The chromosomes may or may not show rearrangements.

Asepsis Without infection or contaminating microorganisms.

Aseptic Technique Procedures used to prevent the introduction of fungi, bacteria, viruses, mycoplasma, or other microorganisms into cell, tissue, and organ cultures. Although these procedures are used to prevent microbial contamination of cultures, they also prevent cross contamination of cell cultures. These procedures may or may not exclude the introduction of infectious molecules.

Axenic Culture A culture without foreign or undesired life forms. An axenic culture may include the purposeful cocultivation of different types of cells, tissues, or organisms.

Callus An unorganized, proliferative mass of differentiated plant cells; a wound response.

Cell Culture Maintenance or cultivation of cells *in vitro,* including culture of single cells. In cell cultures, the cells are no longer organized into tissues.

Cell Hybridization The fusion of two or more dissimilar cells leading to the formation of a synkaryon.

---

1. The terms listed in this appendix are adapted from *In Vitro Cellular and Developmental Biology, 26,* pp. 97–100 (1990). Copyright © 1990 by the Tissue Culture Association, reprinted by permission. Tissue Culture Association Terminology Committee members: Stephen Mueller, Coriell Institute for Medical Research, Camden, NJ; Michael Renfroe, James Madison University, Harrisonburg, VA; Warren I. Schaeffer (Chair), University of Vermont, Burlington, VT; Jerry W. Shay, The University of Texas, Southwestern Medical Center at Dallas, Dallas, TX; James Vaughn, U.S. Department of Agriculture, Beltsville, MD; Martha Wright, CIBA-GEIGY, Research Triangle Park, NC.

Cell Line The product of the first successful subculture of a primary culture. Cultures from a cell line consist of lineages of cells originally present in the primary culture. The terms *finite* or *continuous* are used as an adjective if the status of the culture is known. If not, the term *line* will suffice. The term *continuous line* replaces the term *established line*. In any published description of a culture, one must make every attempt to characterize the culture or describe its history. If such a description has already been published, a reference to the original publication must be made. If a culture is obtained from another laboratory, the proper designation of the culture, as originally named and described, must be maintained and any deviations in cultivation from the original must be reported in any publication.

Cell Strain Strain derived from either a primary culture or a cell line by the selection or cloning of cells having specific properties or markers. In describing a cell strain, its specific features must be defined. The terms *finite* or *continuous* are to be used as an adjective if the status of the culture is known. If not, *strain* will suffice. In any published description of a cell strain, one must make every attempt to characterize the strain or describe its history. If a description has already been published, a reference to the original publication must be made. If a culture is obtained from another laboratory, the proper designation of the culture as originally named and described must be maintained and any deviations in cultivation from the original must be reported in any publication.

Chemically Defined Medium A nutritive solution for culturing cells in which each component is specifiable and, ideally, is of known chemical structure.

Clonal Propagation Asexual reproduction of plants, the results of which are considered to be genetically uniform and originated from a single individual or explant.

Clone In animal cell culture terminology, a population of cells derived from a single cell by mitoses. A clone is not necessarily homogeneous, and, therefore, the terms *clone* and *cloned* do not indicate homogeneity in a cell population, genetic or otherwise. In plant culture terminology, the term may refer to a culture derived as in animal cell culture or to a group of plants propagated by only vegetative and asexual means, all members of which have been derived by repeated propagation from a single individual.

Cloning Efficiency The percentage of cells plated (seeded, inoculated) that form a clone. One must be certain that the colonies formed arose from single cells in order to properly use this term. (See also **Colony Forming Efficiency.**)

Colony Forming Efficiency The percentage of cells plated (seeded, inoculated) that form a colony.

Complementation The ability of two different genetic defects to compensate for one another.

Cryopreservation Ultralow temperature storage of cells, tissues, embryos, or seeds. This storage is usually carried out using temperatures below $-100°C$.

Cybrid The viable cell resulting from the fusion of a cytoplast with a whole cell, thus creating a cytoplasmic hybrid.

Cytoplast The intact cytoplasm remaining following the enucleation of a cell.

Cytoplasmic Hybrid Synonymous with cybrid.

Cytoplasmic Inheritance Inheritance attributable to extranuclear genes; for example, genes in cytoplasmic organelles, such as mitochondria or chloroplasts or in plasmids.

Differentiated Cells that maintain, in culture, all or much of the specialized structure and function typical of the cell type *in vivo*.

Diploid The state of a cell in which all chromosomes, except sex chromosomes, are two in number and are structurally identical with those of the species from which the culture was derived.

Electroporation Creation, by means of an electrical current, of transient pores in the plasmalemma, usually for the purpose of introducing exogenous material, especially DNA, from the medium.

Embryo Culture *In vitro* development or maintenance of isolated mature or immature embryos.

Embryogenesis The process of embryo initiation and development.

Epigenetic Event Any change in a phenotype that does not result from an alteration in DNA sequence. This change may be stable and heritable and includes alteration in DNA methylation, transcriptional activation, translational control, and post-translational modifications.

Epigenetic Variation Phenotypic variability that has a nongenetic basis.

Euploid The state of a cell nucleus that contains exact multiples of the haploid number of chromosomes.

Explant Tissue taken from its original site and transferred to an artificial medium for growth or maintenance.

Explant Culture The maintenance or growth of an explant in culture.

Feeder Layer A layer of cells (usually lethally irradiated for animal cell culture) upon which is cultured a fastidious cell type. (See also **Nurse Culture.**)

Friability The tendency of plant cells to separate from one another.

Gametoclonal Variation Variation in phenotype, either genetic or epigenetic in origin, expressed by gametoclones.

Gametoclone Plants regenerated from cell cultures derived from meiospores, gametes, or gametophytes.

Habituation The acquired ability of a population of cells to grow and divide independently of exogenously supplied growth regulators.

Heterokaryon A cell possessing two or more genetically different nuclei in a common cytoplasm, usually a result of cell-to-cell fusion.

Heteroploid The state of a culture in which the cells possess nuclei containing chromosome numbers other than the diploid number. This term is used to describe only a culture and not individual cells. Thus, a heteroploid culture would be one that contains aneuploid cells.

Homokaryon A cell possessing two or more genetically identical nuclei in a common cytoplasm, resulting from cell-to-cell fusion.

Hybrid Cell The mononucleate cell that results from the fusion of two different cells, leading to the formation of a synkaryon.

Hyperhydricity A condition of plants in cell culture that have abnormal shoot development. The shoots have a glass-like appearance and do not establish well in a potting mix. Generally, this condition is characterized by large intercellular spaces, less epicuticular wax, fewer stomata, chloroplasts with small grana and lacking starch grains. *Vitrification* is another term used to describe this condition.

Induction Initiation of a structure, organ, or process *in vitro*.

In Vitro Propagation Propagation of plants in a controlled, artificial environment, using plastic or glass culture vessels, aseptic techniques, and a defined growing medium.

In Vitro Transformation A heritable change, occurring in cells in culture, intrinsically or from treatment with chemical carcinogens, oncogenic viruses, irradiation, transfection with oncogenes, and so on and leading to altered morphological, antigenic, neoplastic, proliferative, or other properties.

Juvenile A phase in the sexual cycle of a plant characterized by differences in appearance from the adult and the lack of ability to respond to flower-inducing stimuli.

Karyoplast A cell nucleus, obtained from the cell by enucleation, surrounded by a narrow rim of cytoplasm and a plasma membrane.

Line See **Cell Line**

Liposome A closed lipid vesicle surrounding an aqueous interior; may be used to encapsulate exogenous materials for ultimate delivery into cells by fusion with the cell.

Meristem Culture *In vitro* culture of a generally shiny, dome-like structure measuring less than 0.1 mm in length when excised, most often excised from the shoot apex.

Micropropagation *In vitro* clonal propagation of plants from shoot tips or nodal explants, usually with an accelerated proliferation of shoots during subcultures.

Morphogenesis (1) The evolution of a structure from an undifferentiated to a differentiated state. (2) The growth and development of differentiated structures.

Mutant A phenotypic variant resulting from a changed or new gene.

Nurse Culture In the culture of plant cells, the growth of a cell or cells on a contiguous culture of different origin that in turn is in contact with the tissue culture medium. The cultured cell or tissue may be separated from the feeder layer by a porous matrix such as filter paper or membranous filters. (See also **Feeder Layers.**)

Organ Culture The maintenance or growth of organ primordia or the whole or parts of an organ *in vitro* in a way that allows differentiation and preservation of the architecture or function.

Organized Arranged into definite structures.

Organogenesis In plant tissue culture, a process of differentiation by which plant organs are formed *de novo* or from preexisting structures. In developmental biology, this term refers to differentiation of an organ system from stem or precursor cells.

Organotypic Resembling an organ *in vivo* in three-dimensional form or function or both. For example, a rudimentary organ in culture may differentiate in an organotypic manner or a population of dispersed cells may become rearranged into an organotypic structure and may also function in an organotypic manner. This term is not meant to be used with the word *culture* but is meant to be used as a descriptive term.

Passage The transfer or transplantation of cells, with or without dilution, from one culture vessel to another. Any time cells are transferred from one vessel to another, a certain portion of the cells may be lost and, therefore, dilution of cells, whether deliberate or not, may occur. This term is synonymous with *subculture.*

Passage Number The number of times the cells in the culture have been subcultured or passaged. In descriptions of this process, the ratio or dilution of the cells should be stated so that the relative cultural age can be ascertained.

Pathogen-Free Free from specific organisms based on specific tests for the designated organisms.

Plant Tissue Culture The growth or maintenance of plant cells, tissues, organs, or whole plants *in vitro.*

Plating Efficiency This term originally encompassed the terms *attachment (seeding) efficiency, cloning efficiency,* and *colony forming efficiency.* These terms are preferable because *plating* is not sufficiently descriptive.

Population Density The number of cells per unit area or volume of a culture vessel. Also, the number of cells per unit volume of medium in a suspension culture.

Population Doublings The total number of times a cell line or strain's population has doubled since its initiation *in vitro.* A formula for the calculation of population doublings in a single passage is number of population doublings = $3.33 \log_{10} (N/N_0)$, where N = number of cells in the growth vessel at the end of a period of growth and N_0 = number of cells plated in the growth vessel. It is best to use the number of viable cells or number of attached cells for this determination. *Population doublings is synonymous with cumulative population doublings.*

Population Doubling Time The interval during the logarithmic phase of growth in which, for example, 1.0×10^6 cells increase to 2.0×10^6 cells. This term is not synonymous with *cell generation time.*

Primary Culture A culture started from cells, tissues, or organs taken directly from organisms. A primary culture may be regarded as such until it is successfully subcultured for the first time. It then becomes a *cell line.*

Protoplast A plant, bacterial, or fungal cell from which the entire cell wall has been removed. (See **Spheroplast** for comparison.)

Protoplast Fusion Technique in which protoplasts are fused into a single cell.

Pseudodiploid The state of a cell that is diploid but in which, as a result of chromosomal rearrangements, the karyotype is abnormal and linkage relationships may be disrupted.

Reculture The process by which a cell monolayer or a plant explant is transferred, without subdivision, into fresh medium. (See also **Passage.**)

Regeneration In plant cultures, a morphogenetic response to a stimulus that results in the production of organs, embryos, or entire plants.

Shoot Apical Meristem Undifferentiated tissue, located within the shoot tip, generally appearing as a shiny, dome-like structure distal to the youngest leaf primordium and measuring less than 0.1 mm in length when excised.

Shoot Tip (Apex) A structure consisting of the shoot apical meristem plus one to several primordial leaves, usually measuring 0.1–1.0 mm in length; when more mature leaves are included, the structure can measure up to several centimeters in length.

Somaclonal Variation Phenotypic variation, either genetic or epigenetic in origin, displayed among somaclones.

Somaclone Plants derived from any form of cell culture involving the use of somatic plant cells.

Somatic Cell Genetics The study of genetic phenomena of somatic cells. The cells under study are most often cells grown in culture.

Somatic Cell Hybrid The cell or plant resulting from the fusion of animal cells or plant protoplasts, respectively, derived from somatic cells that differ genetically.

Somatic Cell Hybridization The *in vitro* fusion of animal cells or plant protoplasts derived from somatic cells that differ genetically.

Somatic Embryogenesis In plant culture, the process of embryo initiation and development from vegetative or nongametic cells.

Spheroplast A cell from which most of the cell wall has been removed. (See **Protoplast** for comparison.)

Stage I A step in *in vitro* propagation characterized by the establishment of an aseptic tissue culture of a plant.

Stage II A step in *in vitro* plant propagation characterized by the rapid numerical increase of organs or other structures.

Stage III A step in *in vitro* plant propagation characterized by the preparation of the propagule for successful transfer to soil, involving rooting of shoot cuttings, hardening of plants, and initiating the change from the heterotrophic to the autotrophic state.

Stage IV A step in *in vitro* plant propagation characterized by the establishment in soil of a plant derived through tissue culture, either after undergoing a Stage III pretransplant treatment or, in certain species, after the direct transfer of plants from Stage II into soil.

Sterile (1) Without life. (2) Inability of an organism to produce functional gametes.

Strain See **Cell Strain.**

Subculture With plant cultures, the process by which the tissue or explant is first subdivided then transferred into fresh culture medium. (See also **Passage.**)

Substrain The result of isolating a single cell or groups of cells in a strain having properties or markers not shared by all cells of the parent strain.

Suspension Culture A type of culture in which cells, or aggregates of cells, multiply while suspended in liquid medium.

Synkaryon A hybrid cell that results from the fusion of the nuclei it carries.

Tissue Culture The maintenance or growth of tissues *in vitro* in a way that allows differentiation and preservation of their architecture or function or both.

Totipotency The ability of a cell to form all the cell types in the adult organism.

Transfection The transfer, for the purposes of genomic integration, of naked foreign DNA into cells in culture. The traditional use of this term in microbiology implied that the DNA being transferred was derived from a virus. The definition used here describes the general transfer of DNA irrespective of its source. (See also **Transformation.**)

Transformation In plant cell culture, the introduction and stable genomic integration of foreign DNA into a plant cell by any means, resulting in a genetic modification. This is the traditional microbiological definition.

Type I Callus A type of adventive embryogenesis found with gramineous monocotyledons, induced on an explant where the somatic embryos are arrested at the coleptilar or scutellar stage of embryogeny. The embryos are often fused together, especially at the coleorhizal end of the embryo axis. The tissue can be subcultured and maintains this morphology.

Type II Callus A type of adventive embryogenesis found with gramineous monocotyledons, induced on an explant where the somatic embryos are arrested at the globular stage of embryogeny. The globular embryos often arise individually from a common base. The tissue can be subcultured and maintains this morphology.

Undifferentiated With plant cells, existing in a state of cell development characterized by isodiametric cell shape, very little or no vacuole, and a large nucleus and exemplified by the cells in an apical meristem or embryo.

Variant With a culture, exhibiting a stable phenotypic change whether genetic or epigenetic in origin.

Vegetative Propagation Reproduction or plants by a nonsexual process involving the culture of plant parts, such as stem and leaf cuttings.

Virus-Free Free from specified viruses based on tests designed to detect the presence of the organisms in questions.

Vitrification See **Hyperhydricity**.

Index

Note: Page numbers with "f" denote figures; "t" tables.

A

Abscisic acid (ABA), 36, 114
  embryo development *in vitro*, 114–116
Acaricides, 59
Acer pseudoplatanus, 2–3
Activated charcoal, 39
Aerosol fly spray, 59
African violet, 106–108
  explant preparation, 107
  in vitro propagation of, 139–140
  observations, 107
  root tip chromosome squash technique,
    107–108
Agar, 37–38, 83
Agrobacterium, 8, 11, 39
Agrobacterium T-DNA, 6–7
Agrobacterium tumefaciens, 5, 57
Agrobacterium-based transformation, 5,
  155–166
  petunia shoot apex, 160–162
  petunia/tobacco leaf disk, 157–160
  tobacco leaf infiltration, 162–164
Amino acids, 38
  l-forms of, 38
Ampicillin, 55
Androgenesis, 7
Angiosperms, 8, 120
Anther culture, haploid plants from, 103–112
  African violet, 106–108
  anther squash technique, 109
  datura anther culture, 105–106
  tobacco, 108
Antibiotics, 39, 55, 57
Apical meristem, 119–120, 121f
  of infected plants, 82–83
Arabidopsis thaliana, 148
Arborvitae. *See* White cedar
l-Arginine, 38
Aseptic technique, 48–49
  germination of seeds, 49–51
Aseptic transfer area, 24–25
l-Asparagine, 38

Autoclaved glassware, 24
Auxin 2,4-D, 33
Auxin stocks, 33
Axenic culture. *See* Plant cell/tissue culture
Axillary bud-breaking, 9
Azaleas, micropropagation of, 95–97

B

Bacillus macerans, 60
Bacticinerators, 61
Bead sterilizers, 61
Benzyl-benzoate spray, 59
Betula sp., micropropagation of, 98
Biolistics, 156
Bipolar somatic embryos,
  formation of, 3
Birch trees, micropropagation of, 98–99
Boston fern, *in vitro* propagation of,
  129–133
  Stage I, 130–131
  Stage II, 131–132
  Stage III, 132–133
Brassica campestris chloroplasts, 8
Brassica napus protoplasts, 8
Broccoli, explant preparation, 49–50, 65
Bulb scales, 87–89
  garlic propagation from, 123, 124f

C

Cactus, 141–142
Calcium hypochlorite, 54
Callus formation, 1–2
Callus induction, 63–80
  callus initiation, 63–66
  cellular variation from callus
    cultures, 76
  competent cereal cell cultures,
    establishment of, 69–70
  explant orientation, 67–68
  growth curves, 72–75
  salt selection *in vitro*, 70–72
Carbohydrates, 36–37

Printed in the United States
By Bookmasters